Sawing a channel through the ice to allow the sealing ships to leave St. John's harbour. Latter nineteenth century.

Courtesy of Martin R. (Mac) Lee.

Haulin' Rope & Gaff

Songs and Poetry in the History of the Newfoundland Seal Fishery

Shannon Ryan/Larry Small

BREAKWATER BOOKS LIMITED

STUDIES OF THE NEWFOUNDLAND AND LABRADOR FISHERY: HISTORICAL AND CULTURAL SERIES, No. 1, 1978
General Editors: Shannon Ryan and Larry Small

BREAKWATER FOLKLORE AND FOLKLIFE SERIES, No. 1, 1978
General Editor: Kenneth S. Goldstein

Published by
Breakwater Books Limited
617 Empire Avenue
St. John's, Newfoundland
A1E 1X1

Typeset by Breakwater Books Limited

Printed by
The Hunter Rose Company Limited
125 Bermondsey Road
Toronto, Ontario

Cover illustration by Gerry Squires
Ferryland Lighthouse, Newfoundland.

Cover design by Helmut Rath

Canadian Cataloguing in Publication Data

Main entry under title:
Haulin' rope and gaff

(Historical & cultural series; 1) (Folklore & folklife; 1)

Bibliography: p.
Includes index.
ISBN 0-919948-53-7 bd. ISBN 0-919948-52-9 pa.

1. Sealing — Newfoundland — Songs and music.
I. Ryan, Shannon, 1941- II. Small, Larry, 1941- III. Series. IV. Series: Folklore & folklife; 1.

M1977.H8H38 784.4'9'718 C78-001057-4

Dedicated to the Newfoundland sealers,
a rare and hearty breed.

TABLE OF CONTENTS

viii Preface

1 Introduction

The Songs and Poems

13 Wadhams Song
14 The Sealer's Call
15 Captain Henry Thomey
16 We will not go to White Bay with Casey any More
17 Come All ye Jolly Ice-Hunters
18 Old Time Sealer's Song
20 Song of the *John Martin* (1)
21 The *John Martin* (2)
22 Ferryland Sealer
25 The Land of Fish and Seals
26 Driven into Spaniard's Bay
26 Come Down with the Killock
27 Terra Nova Seal Fishing
29 A Newfoundland Hero
32 Captain Bill Ryan left Terry Behind
33 Sealer's Song
35 The Sealing Trip of the *S.S. Greenland* 1891
37 Trinity Bay Tragedy
39 Died on the Ice Fields
41 Loss of the *Maggie*
42 The Wreck of the *Maggie*
44 The Spring of '97
46 The *Greenland* Disaster (1)
48 The *Greenland* Disaster (2)
50 The *Greenland* Disaster (3)
53 Written in Memory of the 48 men lost in *Greenland* Sealing Disaster March 21, 1898 (4)
54 *Greenland* Disaster (5)
55 The *Greenland* Disaster (6)
57 The *Greenland* Disaster (7)
59 The Ice Floes
63 Sealing Fifty Years Ago
64 The Sealer's strike of 1902
65 The Sealer's Gained the Strike
66 The Luck went with the Sealers since Brave Colloway led the strike
67 Ice Bound Hunting Seals
69 The Sealer Lad
70 When the Flippers Strike the Town
71 Arrival of the *Grand Lake* & *Virginia Lake* with Bumper Trips
72 First Arrival — *Aurora* & *Walrus*, Full
73 Arrival of *Aurora*, *Diana*, *Virginia Lake* & *Vanguard*, Loaded
74 The *Nimrods* Song
76 Captains and Ships
79 The Sealer's Song
80 Six Men & one woman taken off the ice at Petty Harbour
81 Success to the Hardy Sealers
82 First arrival from the Seal Fishery
83 Captain Bob Bartlett
85 Ballad of Captain Bob Bartlett
86 Maurice Crotty
88 The Spring of the Wadhams
89 I am a Newfoundlander
91 A Sealer's Love Letter

92 Loss of the *S.S. Algerine*
93 In Memorial
94 The *Newfoundland* Disaster (1)
96 The *Newfoundland* Disaster (2)
98 The *Terra Nova*
99 *Southern Cross*
101 Captain Frederick Harris and The Grates Cove Seal Killers of 1915
102 Au Revoir To our Hardy Sealers
103 The *Kite* Abandoned in White Bay
105 To the Memory of the Late Captain Kennedy
106 The Swiles of Newfoundland
106 Uncle Bill Teller
107 Willy March
108 The Sealers of Newfoundland
109 Success to every Man
110 The Bird Rocks
113 The Cooks of Torbay
114 A Noble fleet of Sealers
116 The Little Swiler
117 Thoughts on the Newfoundland Sealing Voyage
118 The Master Watch
119 *Neptune*, Ruler of the Sea
120 Cotton's Patch (1)
121 Cotton's Patch (2)
122 Full Loads to the Sealers
123 The Sealers' Ball
125 Mrs. Mullowney was three weeks in bed since she ate the fipper stew
126 The Sealing Cruise of the *Lone Flier*
130 The Sinking of the *Newfoundland*
132 The Barrack's Song
133 Hunting Seals
134 The Fisherman's Son to the ice is gone
134 The Girl I left behind me
135 The Sealing Fleet
137 Arctic Ice and Flippers
138 The Old Time Sealers
140 The Old-Time Sealing Fleet
142 Last of the Wooden Walls
144 The Swiler's Song
146 Captain Abram Kean
147 Ballad of Sealing Ships and Sealers
149 The Dying Seal-Hunter
150 The Sealers of Twillingate and New World Island
153 A Sealer's Reply to his Wife
154 Uncle John — The Sealer, 1951
155 We will Always Have Our Sealers
156 Save our Swilers

159 Notes on the Songs
166 Index of Titles and First Lines
173 Bibliography

PREFACE

This collection is an attempt to bring together from various sources songs from Newfoundland's sealing tradition. The collection in not intended to be complete, but is rather an effort to make a beginning towards what one day might result in a more comprehensive work. It is with this in mind that we invite those readers interested in preserving and sharing with others sealing songs which they might know to submit them to us for possible future publication.

In some instances we have included more than one version of the same song. This was done to give readers some understanding of the composition process and the changes which occur when songs and stories are transmitted from one person to another.

We feel there are two reasons why a collection of sealing songs is a necessary publication. In the first place, these songs are historical documents and as such they enable us to learn more about the sealing industry; it is for this reason that the songs have been arranged in a rough chronological order. Second, the songs are also an expression of a fundamental part of our cultural heritage.

Like most other occupations, sealing caught the attention of local composers, foremost being James Murphy (1867-1931) of St. John's who was referred to as the sealer's poet. Moreover, he was not only a writer and composer but also published his own works. Murphy's seven song books contain several selections pertaining to the seal fishery, the most pertinent works being MURPHY'S SEALERS' SONG BOOK (1905) and SONGS SUNG BY OLD TIME SEALERS OF MANY YEARS AGO (1925). These books, which were privately published, are now out-of-print and extremely rare.

We are thankful to James Murphy's son, Michael, for permission to include in our work selections from his father's collection. Michael Murphy himself is also well-known as a local historian and writer with a variety of publications to his credit and others which are forthcoming.
Another local composer and publisher was Johnny Burke (1851-1930). We have included some of his material here and are grateful for the assistance given to us by Paul Mercer of the Arts and Culture Centre Library. Mr. Mercer is known for his book, THE BALLADS OF JOHNNY BURKE (1974), published by the Newfoundland Historical Society. At all times he cooperated with us in discussing Burke as a composer and in seeking out otherwise obscure Burke collections.

We are indebted to John White, among the first and most popular singer of Newfoundland songs, for his kind permission to use selections from his publication, BURKE'S BALLADS, which he found time to compile and edit in the midst of a very busy career.

The Gerald S. Doyle song books, OLD TIME SONGS AND POETRY OF NEWFOUNDLAND, issued in 1927, 1940, 1955 and 1966, have been invaluable throughout our task. The late Gerald S. Doyle (1892-1957), a St. John's businessman, had a keen interest in Newfoundland culture. In his many journeys around the coast by boat he not only distributed school supplies to needy students but also succeeded in collecting a large number of local songs. These songs, along with some taken from other works, he published and distributed free of charge. We are thankful to his wife, Mrs. Gerald S. Doyle, for discussing her husband's work with us, locating a few songs not in the Doyle printed collections, and kindly giving us permission to reprint selections from the Doyle published sources.

The texts of the songs and poems in the collection appear, for the most part, as they did in the source from which they were taken. Where changes have been made it has been with incorrect dates, punctuation, and the italicizing of the names of sealing vessels.

We would also like to thank the following for permission to use their songs or those of their relatives: Arthur Scammell, A.C. Wornell, Otto Kelland, Mrs. Elizabeth Greenleaf, E.J. Bonnell, Gerald Loveridge, Norman Payne, Mrs. Dorothy Cook, Mrs. Michael Colbert, Harry Carter, Mrs. Stella Whelan, Herbert Cranford, Leslie Stansford, John Greene, Nicholas Lane, John Widdowson, Clayton L. King, Pat Byrne, Michael Harrington, F.J. O'Leary, Cyril Byrne, Gerard Patenaude of the National Museum, Tundra Books, Montreal; MacMillan Company of Canada; Susan Sherk and Fred Earle, Memorial University Extension Service and Hal Andrews of the CBC.

Other people have aided us in this undertaking and we should like to take this opportunity to thank them for their help; Burnham Gill and the staff of the Provincial Archives for the use of their photographic collection; the staff of the Newfoundland Studies Center, Memorial University; the staff of the Arts and Culture Centre Library; Jack Martin of ETV; Bertram Shears and Cyril S. Chafe, formerly of Bowring Brothers; Harry Collingwood, Manager, Baine Johnston Limited; Dr. D.G. Pitt, Memorial University; Peter Narvaez, MUNFLA archivist; John Ashton and Martin Lovelace, doctoral students, Department of Folklore, Memorial University; Carole Henderson Carpenter, York University.

We owe a special appreciation to Dr. Kenneth S. Goldstein and Dr. Herbert Halpert of the Memorial University of Newfoundland Folklore Department and Dr. James Tague of the History Department for their advice and encouragement. We are much indebted to Dr. G.M. Story who kindly read the manuscript, gave numerous helpful suggestions, especially with the notes, and who continues to intrigue us with his immense knowledge, often articulated with humour. Heather Wareham deserves our thanks for the map work. Mrs. Susan Snook, Miss Mary Butler, Mrs. Cindy Turpin, Mrs. Sharon Cochrane and Mrs. Pamela Roberts of Memorial University deserve our sincere gratitude for their assistance in typing and proofreading our often illegible scrawl. And our special thanks to Mrs. Ruth Thornley for the musical transcriptions.

Finally, we would like to express our gratitude to those old sealers who helped us acquire the flavour of the seal fishery with their songs and yarns. Among them we would like to mention especially Andy Short, Edward Russell and Joe Ryan of Riverhead, Harbour Grace, and Warwick and Abe Horwood of Moreton's Harbour.

Memorial University of Newfoundland 1978

Shannon Ryan, Department of History

Larry Small, Department of Folklore

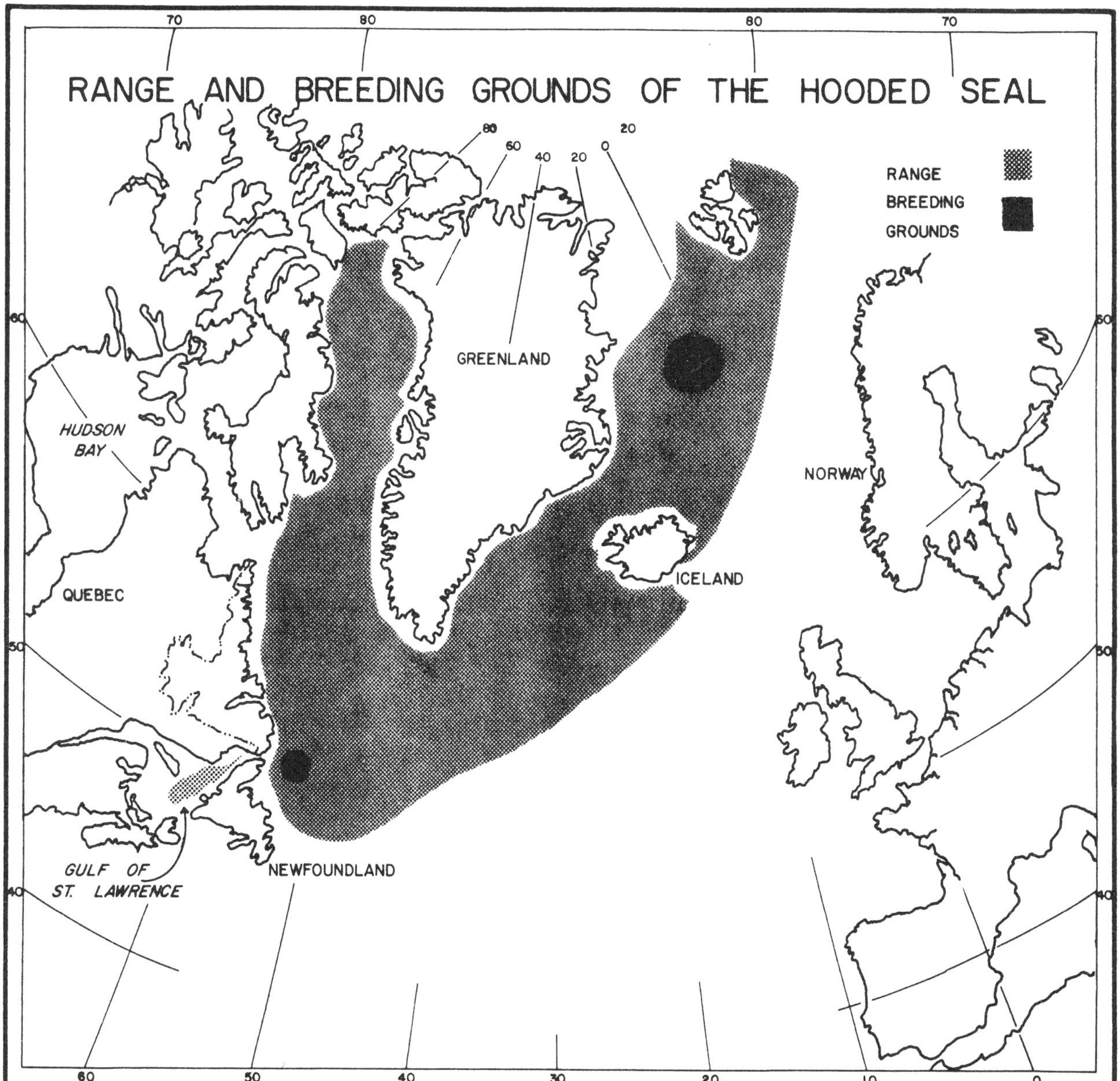
RANGE AND BREEDING GROUNDS OF THE HOODED SEAL
RANGE
BREEDING GROUNDS
GREENLAND
ICELAND
NORWAY
HUDSON BAY
QUEBEC
NEWFOUNDLAND
GULF OF ST. LAWRENCE
70
80
80
70
60
50
40
30
20
10
0

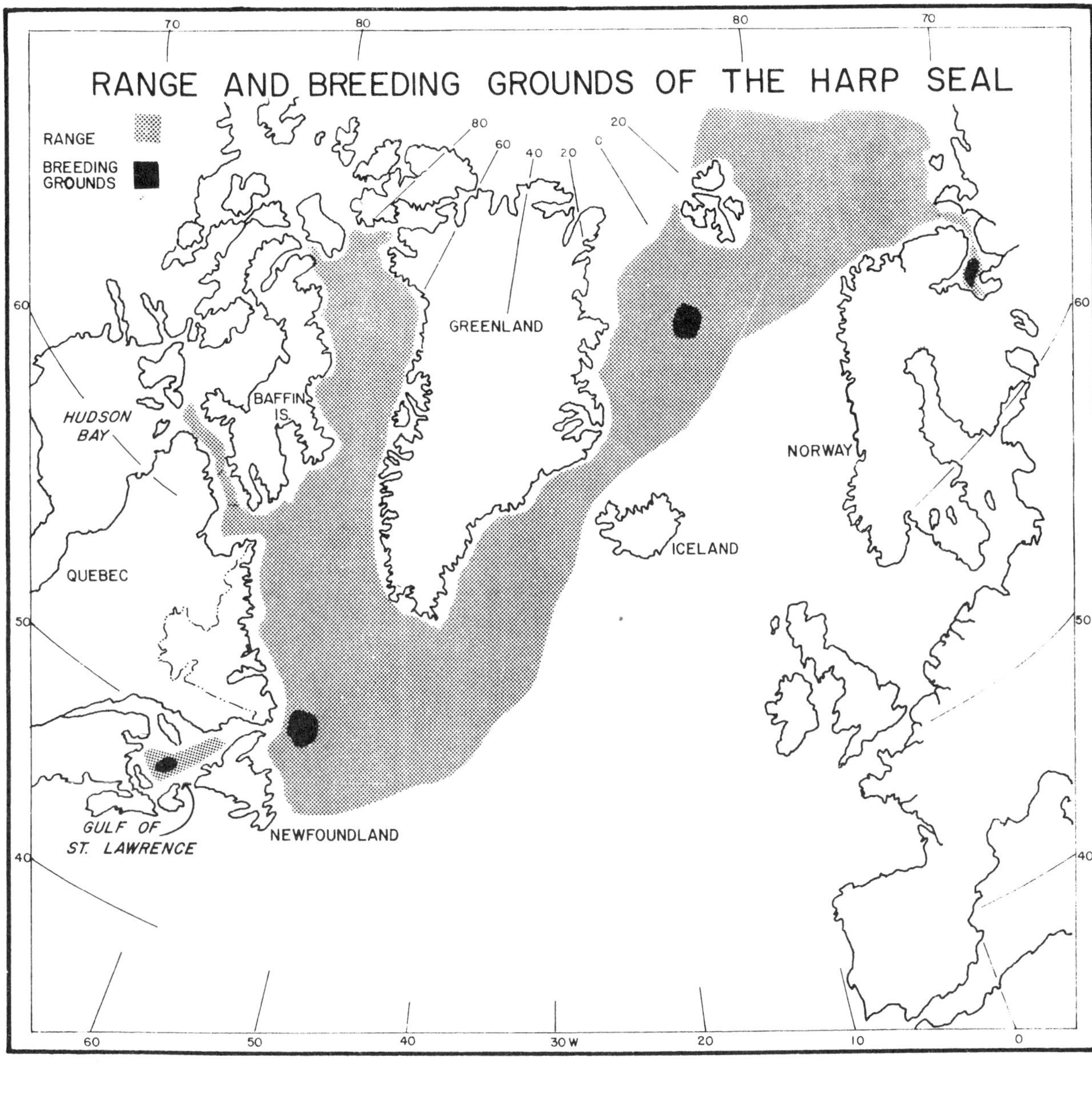
RANGE AND BREEDING GROUNDS OF THE HARP SEAL
RANGE
BREEDING GROUNDS
GREENLAND
BAFFIN IS.
HUDSON BAY
QUEBEC
NORWAY
ICELAND
NEWFOUNDLAND
GULF OF ST. LAWRENCE
30 W

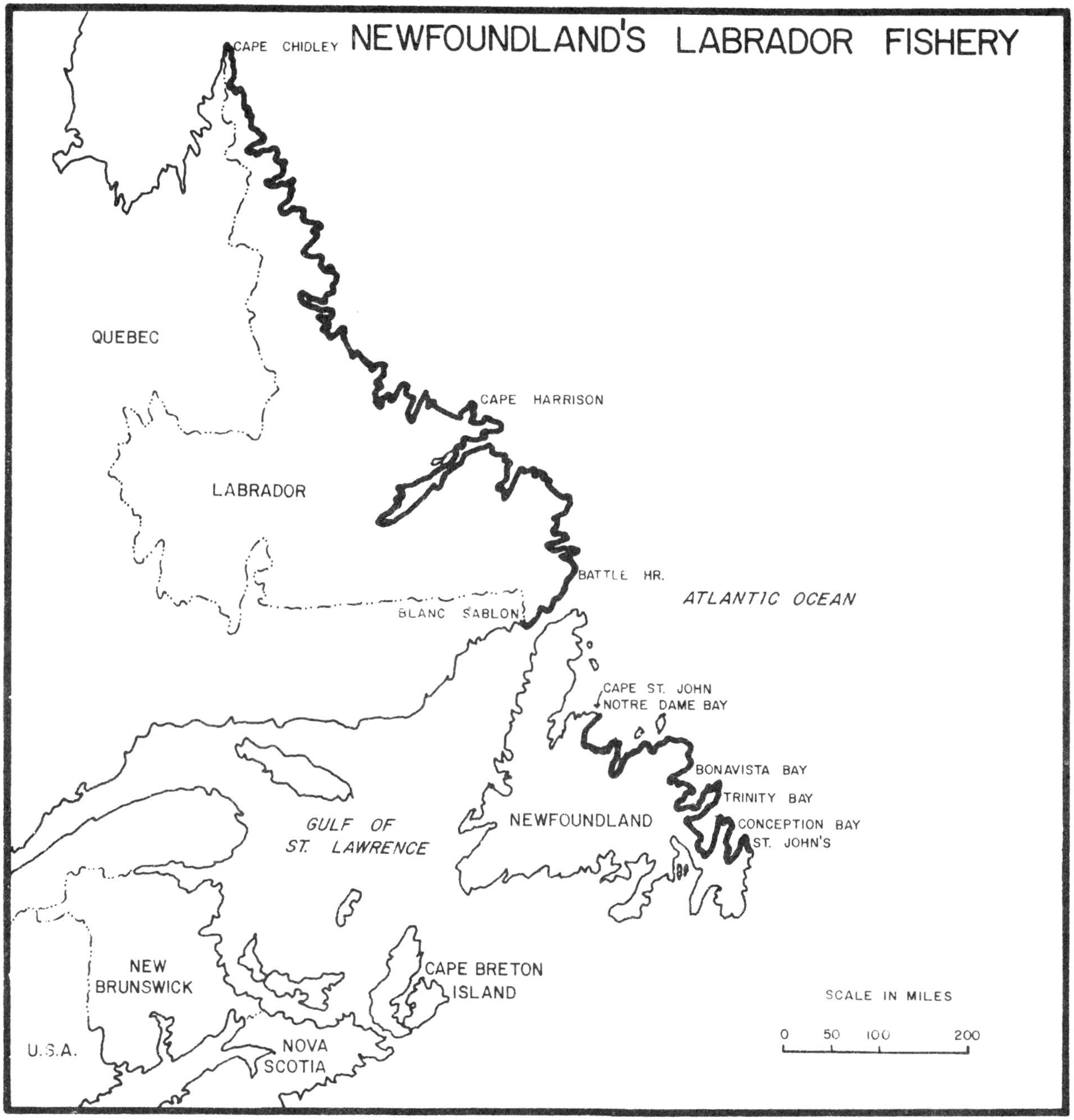
NEWFOUNDLAND'S LABRADOR FISHERY
CAPE CHIDLEY
QUEBEC
CAPE HARRISON
LABRADOR
BATTLE HR.
ATLANTIC OCEAN
BLANC SABLON
CAPE ST. JOHN
NOTRE DAME BAY
BONAVISTA BAY
TRINITY BAY
NEWFOUNDLAND
CONCEPTION BAY
ST. JOHN'S
GULF OF
ST. LAWRENCE
CAPE BRETON
ISLAND
NEW
BRUNSWICK
SCALE IN MILES
0 50 100 200
U.S.A.
NOVA
SCOTIA

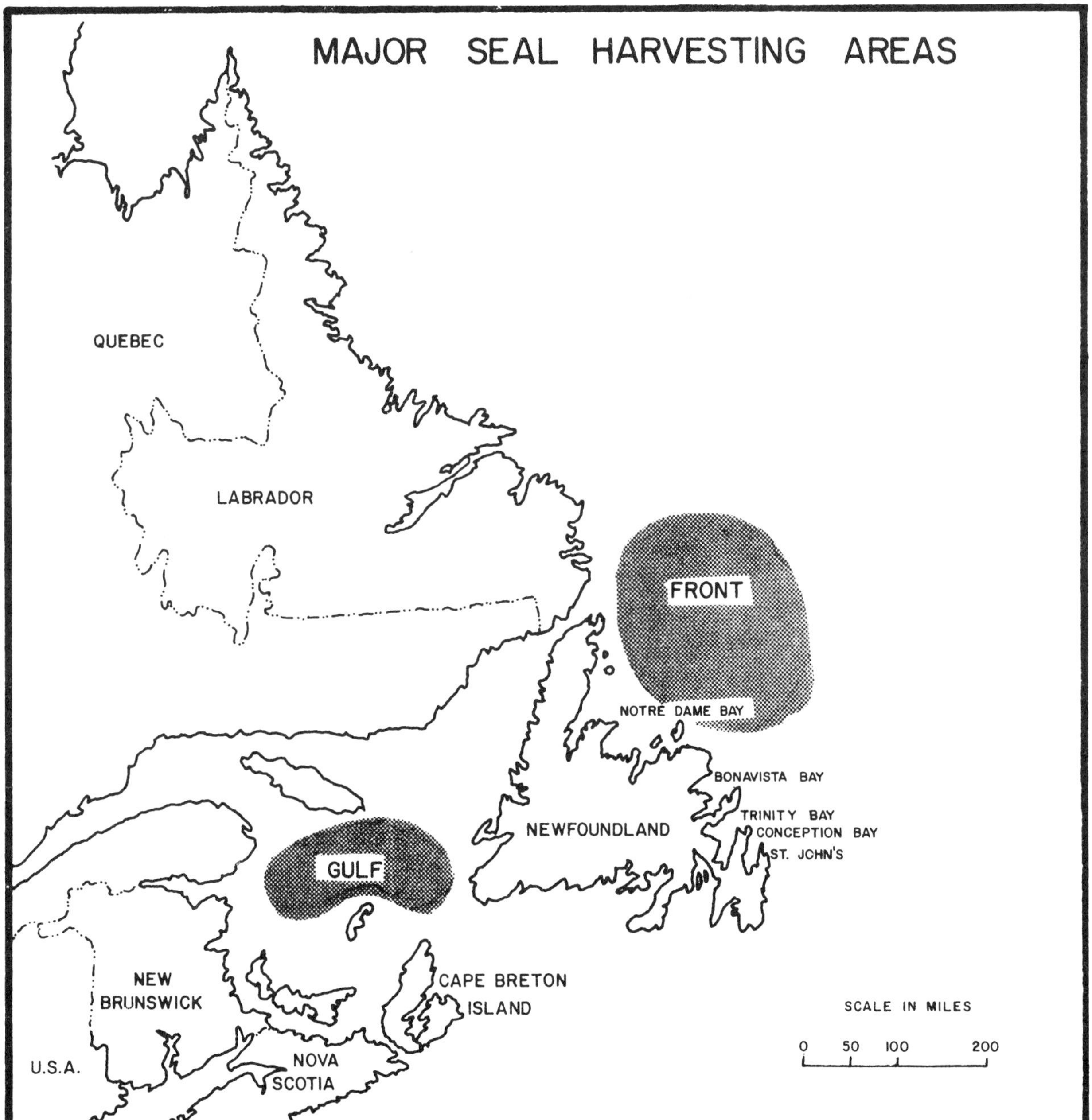
MAJOR SEAL HARVESTING AREAS
QUEBEC
LABRADOR
FRONT
NOTRE DAME BAY
BONAVISTA BAY
TRINITY BAY
CONCEPTION BAY
ST. JOHN'S
NEWFOUNDLAND
GULF
CAPE BRETON
ISLAND
NEW
BRUNSWICK
U.S.A.
NOVA
SCOTIA
SCALE IN MILES
0 50 100 200

INTRODUCTION
The Seal Fishery in Newfoundland's Economy and Culture [1]

The Newfoundland economy and and the Newfoundland culture, in their traditional forms, rested upon and were shaped by the cod and seal fisheries.

The early English settlers on Newfoundland's northeast coast, though originally concerned exclusively with the cod-fishery, learned very quickly to utilize the herds of seals that migrated annually between Newfoundland waters and the Arctic. Nets were used to catch seals in open water, guns used to hunt them, and on occasion the ice floes containing thousands of new-born mammals would drift into the various harbours. Because the adult seals had their young on the drifting ice off the Labrador coast around the end of February the animals were swept southward for several weeks until the young were mature enough to swim. This could create a bonanza as in 1862 when the main herd drifted into Green Bay and it was reported that *"the women and dogs made £10 per'man."* In all these instances seals were harvested by people near their home communities and this type of sealing is still carried on in a few places today. In sealing terminology such sealers have always been referred to as "landsmen" and it was the landsmen who first discovered (as early as the beginning of the eighteenth century) that seal oil was a valuable commodity with a ready market in England.

Newfoundland's population increased very slowly during the first half of the eighteenth century and only somewhat faster during the second half. By the turn of the century, however, the residents found themselves with a monopoly on the European fish trade. This encouraged capital investment and created a demand for labour. Emigrants — particularly Irish — arrived to seek employment especially in the more heavily commercialized centres around St. John's and Conception Bay. High fish prices and plenty of labour encouraged the merchants and planters to outfit fishing ships and send them to the northern coasts of the Island and the Labrador coast for the summer fishery. There they could fish unhindered by neighbors and dry their catch in the vacant harbours and coves. When the war ended, this fishery, which has become known as the Labrador cod fishery, was not totally viable by itself. It was retained, however, for it provided summer employment for ships and men who were

1 *A shorter version of this paper, entitled* "Animals of the Ice: The Seal in Newfoundland Culture", was presented by the author to the Annual Meeting of the American Folklore Society in Philadelphia on November 12, 1976.

engaged in another activity that also originated during the wars — the annual spring seal fishery.

In the 1790s a few ships from Conception Bay and St. John's were fitted out to go and look for seals. They proved to be successful. Each succeeding year the men hunted farther and farther afield acquiring more experience and knowledge all the time. In a very short while they discovered that the seals gave birth around the end of February and that these young could be slaughtered efficiently and easily. The young seals also provided a finer quality oil. Armed with long iron-tipped sticks called gaffs the men swarmed over the icefloes during the early weeks of March killing whitecoats and other species or types of seal they found. The young harps became the mainstay of the industry and the discovery of the breeding patches was the aim of every sealer. The industry grew rapidly and in 1844 roughly 700,000 seals, worth about £330,000, were taken. This represented over one-third of Newfoundland's whole export economy and made up a substantial part of the fisherman's income. After the 1860s the industry began to decline because of over harvesting; this decline was accelerated after 1880.

Once established in the early 1800s the ship seal fishery provided extensive employment; investment in the industry reached its peak in 1857 when reportedly there were 370 vessels and 13,600 men employed. In the 1860s steam power was introduced and the sailing fleet rapidly declined. At the same time the outports declined, for they could not afford steamers, especially since sealing was becoming less and less lucrative. These structural changes were one of the factors that led to strained relations between St. John's and the outports which have not completely disappeared. In any case, by 1895 St. John's owned the entire steam fleet and Conception Bay which had, at one time, controlled about two-thirds of the industry was experiencing massive emigration and economic depression. Beginning in 1906 iron clad vessels were acquired for sealing and from then until 1914 Newfoundland probably had the greatest fleet of ice breakers in the world along with skilled men to operate them. Both ships and men were employed by numerous international Arctic expeditions around the turn of the century. One vessel, the *Terra Nova*, was acquired by Scott on his last expedition to the Antarctic and when Admiral Peary organized his all-American expedition to the North Pole in 1908-09 he hired a Newfoundland sealing captain and crew. The industry, however, could not support the iron clads and they were disposed of during World War I. During the 1920s eight or ten old wooden walls continued to operate with a total of about 2,000 men who by now were earning much less than their predecessors. Nonetheless, in a socio-economic system that operated largely on credit the seal fishery was one of the few sources of cash available to the fishermen. The decline continued into the 1930s and during the 1940s the last old steamer was withdrawn. At the present time one or two motor vessels and less than a hundred men are engaged in this type seal fishery along with a number of landsmen. It continues, however, to provide an important income supplement for a number of fishermen.

Sealing or swilin' as it was called was a rugged, dangerous occupation, new to the experience of the Newfoundlander in 1800, for which certain technological adaptations had to be made. In the beginning when the hunt took place in April and May long barrelled muzzle-loading guns were used. Although this aspect of the hunt became much less important after the discovery of the breeding grounds it continued as part of the industry. Shooting seals was a very wasteful operation for a large proportion of wounded and dead seals were lost. When the sealers began to utilize the breeding patches they quickly adopted the common boat hook into a useful tool. The boat hook, traditionally associated with boats, is a long staff with a metal point and hook at one end. It can be used for a variety of purposes, such as pushing one's boat away from the wharf, pulling one's boat closer to the wharf, and for pushing and pulling objects in the water. The gaff, however, was a staff about seven feet long tipped with a much heavier hook and spike. It could be used to maintain one's balance when running across loose, hazardous ice — a practice that was called copying. It could also be used to pull a companion from the water, to pull a seal pelt or any object closer and to push and pull pieces of ice in place in order to create a temporary bridge. Shavings from a gaff were even used to start a fire during an emergency and there are cases on record when gaffs were burned to keep up steam as the ship tried to reach port at the end of the hunt. But its main purpose was for killing seals, both young and old. It was the emblem of the seal fishery and the badge of manhood. The seal fishery was an adult male activity and a boys's first gaff along with his first trip to the ice was an initiation into manhood, although it seems very likely that this attitude became more prevalent as the seal fishery became more restricted. In any case a boy's father would generally make his first gaff during the winter preceding the hunt. This would provide a topic of conversation on numerous occasions as the father would probably be coaxed to get on with the task. Then there would be discussion as the work continued and finally discussion about the final product. Another essential item which had to be provided before hand was the tow rope which was used for towing the seals to the ship or to a central location. When not in use it was wrapped around the chest under one arm while on the ice so if a man fell in the water he could be saved by a colleague hooking his gaff in the victim's tow rope. And finally the knife and steel which were purchased for the purpose. The knife, different from those used in the cod fishery — the cut-throat and the splitting knife — had a wide, slightly curved blade about six inches long. It was kept exceptionally sharp and worn in a hand-made wooden sheath making it easier to remove or replace when one's hands were cold or wet with blood. These were the major personal items used in the hunt. The ship provided the necessary guns (in later years gaffs as well). In addition, one always carried a wooden box to the ice, referred to as a sealer's chest. Into this went extra clothing, oilskins and various other items such as a checkerboard, tobacco and playing cards. Also, one used it to store and keep various articles of food, like hard bread or a tub of butter, that could be brought back home.

Ashore, when the ships unloaded, the seal pelts went to the seal skinners who removed the two-three inch thick layer of fat from the skin for a fixed price per pelt. In the 1920s there were five rates ranging from five cents per pelt for a young harp to thirty cents for an old hood. Twenty-five seal skinners employed by Newfoundland's largest firm could skin on an average 7,000 young harps a day. A Mr. George Cook established a record of 656 young harps in one nine-hour day. Seal skinning was a highly skilled job and required years of apprenticeship before one became proficient at it.

By the early years of this century the skinners' union restricted membership with new members being admitted only upon the death or retirement of an older member. Their specially-designed instrument was the seal skinning knife with a broad, curved blade about eight inches long. Each man stood behind a table that sloped away from him. He held the skin in one hand and cut the fat away so that it slid off the table and into a container on the other side. The skin and the table had to be free of any foreign objects such as pieces of ice or fat, for either of these could raise the skin into the path of the knife resulting in a cut in the skin and during the 1920s the skinners were charged ten cents a cut. Needless to say, seal skinning was a seasonal occupation and skinners worked at other jobs during the rest of the year. In 1925 machinery was introduced to perform this task and seal skinning very quickly ceased to be a local craft.

Another feature of the seal fishery is that it developed, towards its later years, a highly differentiated crew structure along with a terminology to define this structure. In the beginning small vessels and crews were used but it was found that larger operations were more economical. Consequently by the end of the nineteenth century crews of over 300 men were not uncommon and this demanded a much greater degree of organization so that these large numbers could be fully utilized. The captain was in over-all command and he was assisted by the second hand. The crew was divided up into four watches each under a master watch. Each watch contained a 'scunner' who stood in the barrel on the forward mast and directed the ship through the ice and a 'bridgemaster' who relayed the instructions on to the 'wheelsman' and his crew who were all members of the same watch. A 'quarter master' was stationed astern to shout directions whenever the ship was obliged to go astern. The 'barrelman' occupied the barrel on the after mast and it was his job to search for seals with his spy glass. To take charge of a watch or become a master watch a man needed wide experience in the seal fishery as well as intelligence and courage and later he needed a certificate indicating he was qualified for this position. He was responsible for assigning the various tasks to his whole watch on and off the ice. He appointed the scunner, the wheelsman and others to various positions on the ship, arranged for his watch to shift coal and load and stow away the seals and led his men on the ice. He appointed 'ice masters' in charge of each small group of men on the ice and it was their responsibility to look after their men who in turn had to obey their instructions. As the job situation tightened up around the turn of the century the master watch's position became more prestigious, as did all positions of authority, for there were more qualified men than there were openings and the captain of one ship might be forced to become a second hand on another ship, while a second hand might be forced to become a master watch and a master watch to seek employment as a regular sealer. The ship also carried guns and the 'gunners' were chosen by the captain. Each gunner could choose his own men to carry the cartridges. These men were called 'dogs' and they were also required to cut off the tails of the seals shot by the gunner and bring them back to the captain as evidence of the kill. If a breeding patch could not be found the gunners searched the loose ice for old seals — a process that was referred to as 'swatching', probably an abbreviation of the words swile watching. However, when not so employed or when young seals were spotted the gunner and the dogs reverted back to their positions as regular sealers and everybody who carried a gaff and killed seals was referred to as a 'batsman' while engaged in that activity on the ice.

Not only did the job become more specialized, but as men became more familiar with the seals they began to make distinctions, sometimes very subtle ones, between the different types of seal, particularly of the harp family. First and foremost there was the young harp which was white or cream and referred to as a whitecoat. This type made up the vast bulk of the harvest. The whitecoats have an innocent expression and a very human-like plaintive cry which often disconcerted some sealers, particularly the younger men. When the whitecoat began to lose its fur it was referred to as a 'raggedy jacket' and when it took to the water at several weeks of age it became known as a 'beater'. If a whitecoat was born dead it was known as a 'cat' and would not lose its fur. This type of seal became the property of the finder and was skinned at the sealer's leisure. It was then sold to a dealer in port who had it tanned and stuffed for sale to the public. Skinning a cat properly was a specialized feat and one of the leisure activities. A small hole was cut in the stomach and the whole carcess was removed through it. Very young whitecoats were also referred to as cats. In the beginning, seals were sold by number and type but in 1835 a sealing captain in St. John's sold a load of cats as regular whitecoats without the buyer seeing what he had bought and after that they were all bought by weight. A whitecoat that had been abandoned by its mother became very thin and thereby valueless and kept up a continuous cry until its death or until it could take to the water and fend for itself. This type of whitecoat was referred to as a 'screecher'. A young harp of one or two years old was referred to as a 'rusty jacket' or a 'rusty' because of the brownish spots on its coat. When these disappeared it was called a 'bedlamer' and when it reached maturity it was referred to as a 'dog' if male or a 'bitch' if female. However, a female that had just had its first pup was referred to as a 'black bitch' or a 'smutty bitch' because of its darker color. Similar designations were used to distinguish the various types of hood seals. The most famous hood seal in both song and story is the old 'dog hood'. It has a leather-like bag over its head which expands when it senses danger and which protects the head from blows. Only its throat is vulnerable to a batsman and although one batsman can kill a dog hood two are preferable — one man to distract the animal while the other hits it across the throat. They often weigh 800 pounds and are dangerous. Despite these various distinctions, seals were generally sold to the firms in one of five categories: young harps, old harps, bedlamers, young hoods and old hoods.

The seal fishery also developed unique features in terms of medicine and sickness. Ice blindness was a common danger and before and even after colored glasses became available many men avoided that affliction by smearing their cheeks with blood or by tying a strip of seal meat across the bridge of the nose and under the eyes. A sealer thus masked must have presented a rather unusual appearance. Cuts and bruises appear to have been prone to infection and it was every man's dread that he would get a swile hand or a swile finger. The fingers could become shriveled in appearance and stiff and hook-shaped. All sealers felt that it could happen as a result of a cut, but more recent evidence seems to indicate that it probably resulted from strain. Furthermore, many sealers had never seen any one with a swile finger so that this widespread fear was probably rather irrational.

In the early days there was no spare time as men laboured with oar and sail to propel their crafts through the ice floes. However, with the advent of steam there were periods of inactivity: on the way to the ice, while the ship was searching for seals, during periods when the ship might be jammed in the ice with no seals in sight, and on the return voyage. One job that had to be attended to was the shifting of coal to make room for the seals, but apart from that and some lesser duties the men had a fair amount of leisure time on their hands. They played checkers and cards, some sang songs or told stories and jokes, others read books and magazines; the occasional sealer would have a 'cat' to skin and a number of men engaged in wood carving, particularly the carving of wooden buckets. A man would take a stick of wood, two inches thick and between eight and twenty-four inches long, and carve a hanging bucket on each end. The final result was a short stick with a bucket hanging loosely from each end, all in the once piece. Sometimes decorations were carved on them and they were brought home to hang as ornaments in the sealer's kitchen or parlour. In some cases they were put on stands and made even more elaborate. This was an attractive leisure activity for it did not require much wood nor special tools. A piece of broken gaff provided the raw material, the sealer's knife was the tool, and the carver's talents and patience did the rest.

The sealer had certain personal possessions which were constant reminders of the seal fishery to his family and the community at large. While his sealing chest would generally be put our of sight, his tools could be seen in his store loft by those who visited him. Also, the wooden buckets hanging in his kitchen were constantly on public display. However, affinity between the sealing and non-sealing members of the society was further reinforced by specific foodways and the expressive culture. Songs and stories served to bring the men's experiences at the seal fishery to the community. The seal flippers that the men brought home became one of the most famous delicacies in Newfoundland and recipes for this dish added to the oral culture.

The biggest connection between the seal fishery and the non-sealing element of the society involved the cooking and eating of seal flippers as a delicacy on the island's northeast coast. From the beginning of the industry sealers made a habit of leaving one flipper attached to each pelt. The flippers, sometimes pronounced 'fippers', were sold locally when the ships returned to port and the money was shared among the men. This annual sale of flippers took on the air of a festival: the sealers selling their goods and drinking rum to make up for a long abstention. The docks would be lined with buyers, and with men, women, children and dogs just looking on. Even to this day a remnant of the annual flipper market remains in St. John's.

Seal meat is very oily and the problem has to be taken care of before the flipper is cooked. Consequently recipes for seal flipper are generally passed from one to another more carefully than most other recipes. This is so because improperly cooked flippers can be inedible. Moreover, since they are rare and seasonal, people cannot, generally, allow themselves the luxury of experimentation. Although they are transmitted orally, seal flipper recipes occasionally find their way into Newfoundland cookbooks. Six cookbooks published within the last twenty years contain four seal recipes — all of them very similar. One could expect a greater emphasis but seal is such a seasonal dish that cookbook publishers might be inclined to overlook it. The first step in cooking seal flippers is removing all the fat and the reader is not always advised to do this although it is crucial. The common method, after the obvious fat is removed, is to soak the flippers in water and baking soda. This brings out a distinctive coloration in the fat and it can be readily indentified and then cut away with a sharp knife or a razor blade. Another method is to parboil the flippers before cooking and a method which is becoming more popular is to remove the outer fat, freeze the flippers and then dip them in boiling water. When removed from the water the fat can be scraped away from the frozen meat. In any case, the removal of all fat is essential. When this has been done the flippers are cooked with salt pork by a combination of frying, simmering and baking. Vegetables are usually cooked in the same pot and in most cases a pastry is put over the entire dish. With proper care and attention the final product is quite delicious. Because of its scarcity, the care needed in preparing it, and the superb quality of the final result, the seal flipper had an important place in traditional culture.

As one would expect, the seal fishery became the inspiration for various song writers. Of the songs collected by L.G. Small and myself the earliest record of publication is in 1833. This collection includes some songs that were never published but no doubt neglects to include a fair number that were published in magazines, newspapers, booklets and broad sheets. Nevertheless, it is a representative collection and from it one can draw certain conclusions.

It is impossible to identify the composers of all these songs, but some information is available. We can identify some songs that were written by sealers and others that were written by non-sealers. One Scottish woman in St. John's in the 1850s described Newfoundland in glowing terms as the land of fish and seals. A Newfoundland woman wrote a song commemorating the Southern Cross disaster 1914. Some were written by a newspaper publisher, several by popular local poets. Songs written by sealers give more intimate details about the sealing voyage, deal with specific types and are written in the first person. Minor incidents are described as are the actions and reactions of various members of the crew, e.g.

The skipper he came forward
And he swore once or twice
Took the kettle by the hangers
And threw it on the ice
Oh! I never felt so sorry
Since the day that I was born
When I saw my little piper
And it floating down astern

In addition, these songs tend to use terminology which is, in many cases, unfamiliar to non-sealers, words like 'cat', 'scunner', and 'swatch'. Songs written by non-sealers are very often written in the third person and occasionally use the second person. Many are very general, e.g.

The stormy month of March is come, boys,
  muster and prepare
To leave your home o'er seas to roam
  and take the sealers' fare

or

We boast not of our mighty dead,
  but of our living brave
Who the frozen ocean fearless tread,
  or skim the angry wave
Heaven's blessing on their gallant hearts
  our country's pride and weal
Who brave the perils of the deep
  in quest of fish and seal.

When they deal with specific topics there is a tendency to deal with those that have an impact on the people at home, particularly a tragic impact: the loss of the *Southern Cross* for example when 173 sealers were lost. And it is significant that a Newfoundland woman should have written this song, for tragedy affected the women and children most of all.

From what information is available it appears that an indepth study of the authorship of sealing songs would provide additional insights into the seal fishery and its place in the culture. Not knowing the composer, of course, is one of the characteristics sometimes associated with traditional song.

Not only has information about the authorship of most sealing songs been lost but so also has the date of composition or publication. However, with a few dates and other information we can discuss the songs in their historical context. Our earliest songs describe events involving small sailing ships with small crews, while our later songs describe large steamers with large crews. In 1830 the average crew contained twenty-one men and in a song which was published in 1833 we hear about the adventures of a sailing ship with twenty-eight men. In the song about the loss of the *Southern Cross* in 1914 we hear about the loss of a steamer and one hundred and seventy men. Other changes in the seal fishery are reflected in the songs: in the 1830s and 1840s the words 'ice hunters' and 'fishermen' are used to describe the men who went to the ice, while later songs refer to them as sealers. Also the four oldest songs in our collection have their setting in the outports which controlled most of the industry at the time. In addition the severe competition and callousness that was present in the industry by 1900 resulted in a number of tragedies that are documented by songs.

Nevertheless, although they vary in quality, the songs treat the seal fishery in a matter-of-fact way. They reflect the concern of people for the dangers but they contain no suggestion of bitterness towards the captain and owners. Extreme hardship is accepted and, if necessary, so is death. The songs indicate that the sealers depended on their own self-reliance, believing firmly that God would look after them when all else failed.

There developed an element of machismo that was actively encouraged by the captains and owners as the industry became restricted and profits declined. Was a fisherman 'man' enough to be a sealer? This was a far cry from the 1840s when fishermen literally went sealing as a seasonal fishery occupation. Three strikes in 1842, 1860 and 1902 were aimed at eliminating berth money — the price a sealer paid to go to the ice — and were not concerned with living and working conditions. When Johnny Burke, a very individualistic song writer around the turn of the century, praised the sealers for striking in 1902 he did not condemn the industry but praised the settlement. Some songs that refer to people by name were probably written and sung in order to gain favour or get revenge. However, all songs from the comic to the tragic were sung on festive occasions in the community. Although their major function was to entertain they also served to educate non-sealers about sealing.[2]

In all, the seal fishery was part and parcel of the Newfoundland economy and culture. In both the expressive and material culture its influence spread throughout the

2 It should be noted here that personal experience narratives were very common and their function was similar to that of songs. However, they were much more prevalent and played a greater role in both education and entertainment.

whole island. Even the government took advantage of this and issued at least eight postal stamps illustrating seals or the seal fishery between 1866 and 1937. However, when the industry collapsed its place in the culture began to decline rapidly for it had few general applications in terms of the broader Newfoundland tradition. Today, most of what few seals are taken, are taken by landsmen in motor boats along the northern coast and the rest are taken by one or two small motor vessels that operate in the traditional manner except that they are required by Canadian law to use the Norwegian hakapik instead of the traditional gaff.

It is to be hoped that the collection of songs which follows will help to preserve a part of Newfoundland's traditional culture and explain her past not only because of their intrinsic value as a major part of the expressive culture but also because of their significance as historical documents.

Shannon Ryan

A sketch of Harbour Grace, one of the most important of the sealing ports, in its prime. c. 1850.

Courtesy of Martin R. (Mac) Lee.

Wadhams Song

From Bonavista Cape to the Cabot Isles
The course is north full forty miles,
When you must steer away North East
Till Cape Freels, Gull Island bears West North West.

Then North, North West thirty-three miles,
Three leagues off shore lies Wadhams Isles;
Where of a rock you must beware,
Two miles sou' South East from off Isle bears.

Then North West by West twelve miles or more,
There lies Round Head on Fogo Shore;
But Nor' Nor West, seven or eight miles
Lies a sunken rock near the Barrack Isles.

Therefore, my friend, I would you advise,
Since all these rocks in danger lies;
That you may never amongst them fall,
But keep your luff and weather them all.

As you draw near to Fogo Land,
You'll have fifteen fathoms in the sounding sand;
From fifteen to eighteen, never more,
And that you'll have close to the shore.

When you abreast of Round Head be,
Then Joe Batt's Point you'll plainly see;
To starboard then three or four miles,
You'll see a parcel of damned rugged Isles.

When Joe Batt's Arm you are abreast,
Then Fogo Harbour bears due West;
But untold Fortune, unlucky laid,
A sunken rock right in the trade.

So West Nor' West you are to steer,
Till Brimstone Head doth plain appear;
Which over Pelley's Point you'll see,
Then of that danger you are free.

And as you draw within a mile,
You'll see a house on Syme's Isle;
The mouth of the channel is not very wide,
But the deepest water is on the larboard side.

When within Syme's Point you have shot,
Then three fathoms of water you have got;
Port hard your helm and take care,
In the mid channel for to steer.

When Pelley's Point you are abreast,
Starboard haul and steer Sou' Sou' West,
Till Pelley's Point covers Syme's stage,
Then you are clear I will engage.

"Wadhams Song" is called after the author who wrote it in the year 1756. It was placed on record in the Admiralty Court in London, shortly after it was composed, and was considered the best coasting guide in those days for that part of our Island Home to which it refers.
(Original note)

Reprinted with permission from James Murphy, SONGS SUNG BY OLD TIME SEALERS OF MANY YEARS AGO. St. John's: James Murphy, (1925), 15.

The Sealer's Call

I must go up to the ice again,
To the fields of purest white,
To the giant icebergs' northern home,
To the fringe of Arctic night.
Up to the storms of frost and snow
Where the north king reigns supreme,
Where the waves lash high with the demon rage
And the winter's winds are keen.

I must go up to the ice again,
To the home of the harp and hood,
To the Whitecoat's nest on the icy floe,
Up where the air is food.
What, though my head be white with age,
The call comes strong and loud;
I crave for the sight of the baby seal,
In its nest on the icy mound.

I must go up to the ice again
And mix with a crew of men,
Fearless and bold on the icy floe,
Willing and zealous and keen,
Men who can leap from pan to pan,
While the surging ice-fields roll,
Men undaunted and unafraid
Of the blizzards from Arctic Pole.

I must go up to the ice again,
Though comforts are unknown,
Where the pay is small and the work is hard
And the winds cut to the bone.
I long to feel the touch of the gaff
On the nose of the baby seal,
To feel the rope of my shoulder strain
With a tow that makes me reel.

I must go up to the ice again,
On one of the sturdy ships,
I love to feel her butting the ice
While she shivers, rises and slips.
I want to watch her nearing the pans,
Where the pelts are put in a heap,
To hear the twirl of the donkey winch
And with my comrades leap.

I must go up to the ice again,
The call comes strong to me,
The call of the north is in my blood
I yearn for the frozen sea.
Then when I'm nearing journey's end
And death doth round me prowl
I trust it will be on a stormy day
When the northern tempests howl.

Reprinted with permission from Solomon Samson, A GLIMPSE OF NEWFOUNDLAND IN POETRY AND PICTURES. Edited by Robert Saunders. (Poole, England: J. Looker Ltd., printers, 1959), 37.

Captain Henry Thomey

Upon the past I'm thinking,
To it my heart is linking
When fifteen thousand hardy men
Trod the frozen floe.
Oh, those old days were merry
And everyone felt cheery
When men sailed 'long with Terry
And Thomey years ago.

The Thomeys were noted seal killers during the early years of the nineteenth century. They were natives of Mosquito but later they moved to Harbour Grace. (Eds.)

Reprinted with permission from James Murphy, THE SEAL FISHERY. (St. John's, 1916.) 41.

We Will Not Go To White Bay With Casey Any More

Tom Casey being commander
Of the *St. Patrick* called by name.
With 28 as brave a boys
As ever ploughed the main.
It was upon the 1st of March
From Carbonear we set sail,
Wind from the West South West, my boys
A smart and a pleasant gale.

We sailed by Cape St. Francis
We kept it well in view,
It was early day next morning
When we were off Baccalieu.
While sailing down by Baccalieu
Five whitecoats we took in.
Cheer up, my boys, "Tom" Casey cries,
Five thousand we'll bring in.

This gave us great encouragement,
We all joined heart and hand
Like Trojans bold and undaunted
For to poke her through the jam.
The poor man was mistaken
Upon that very day
Which left us jammed in White Bay
Until the last of May.

Now we are clear of White Bay
And home is our intent,
To see our wives and sweethearts
That we left behind lament.
The wind it came ahead of us
And our course we could not lie,
'Twas in Stag Harbour Tickle, boys,
We warped her high and dry.

We lay there six long days, my boys,
Our course got very low,
Some got hungry, some got dry,
And from her we did go,
And when we got to Pacquet
Relief was there to find,
And it was consolation
For the hungry and the blind.

It was of one John Dooley,
May the Lord prolong his days,
And while he lives on the French Shore
May he live at his ease;
May he be as rich as Dives,
Or rather three times more,
He is a proper gentlemen
That lives on the French Shore.

And there is one thing more, my boys,
Which I ne'er shall forget,
His wife is the brightest woman
That ever I saw yet.
He is open hearted,
His charity to bestow,
I will have his name recorded
Wherever I shall go.

Now eight men we left behind us,
Provisions we had none,
Along with Captain Kelly
In the *Kitty* we came home.
And now we're clear of White Bay
And landed safe on shore,
We will not go to White Bay
With Casey any more.

This is a very old song, written in the early days of the sailing vessels.
(Original note)

Reprinted with permission from James Murphy, SONGS SUNG BY OLD TIME SEALERS OF MANY YEARS AGO. (St. John's: James Murphy, 1925.), 3.

Come All Ye Jolly Ice-Hunters

Come all you jolly ice-hunters and listen to my song,
I hope I won't offend you, I don't mean to keep you long;
'Tis concerning an ice-hunter from Tilton Harbour sailed away,
On the fourteenth day of March, eighteen hundred and thirty-three.

William Burke was our commander, the *Daniel O'Connell* was our good ship's name.
We had twenty-eight as smart a lads as ever crossed the main;
As off with flying colours to the Northward we did steer,
So mark what followed after, to you I will declare.

'Twas on the fourth of April, right well I mind the day;
About four o'clock in the evening our tow line gave away;
The wind came from the North-west and bitterly did blow,
Our Captain cries "Stand by me boys, out of the ice we'll have to go."

"Stand by your topsail halyards, stand by to let them go,
Be quick, I say, and make no delay, your topsail clear also;"
He watched his opportunity, and soon he had her free,
Saying, "God bless the "Brave *O'Connell*" see how she stems the sea."

At six o'clock next morning, we were a dreadful wreck,
Our topmast went overboard, about three feet from the deck;
In this perilous condition for two long days we lay,
So we left her to God's mercy, and to the raging sea.

We could not keep a light below, the seas ran mountains high.
And expecting every minute that we were doomed to die;
At eight o'clock next morning, all hands were called on deck,
Some to rig up jury masts, and more to clear the wreck.

Now in a few days after assistance was at hand,
At six o'clock in the morning the watch espied the land;
So now thanks unto Providence we're safe on shore at last,
We'll drink to one another and drown sorrow in the glass.

This song was written in 1833. It is about the oldest song of a sealing nature now in existence, and has "brought down the house" in the for' castle of many a sealer in the days of the Square Riggers.
(Original note)

Reprinted with permission from Gerald S. Doyle, OLD TIME SONGS AND POETRY OF NEWFOUNDLAND. (St. John's: Gerald S. Doyle Ltd., 1927), 15.

Old Time Sealers' Song

We'll sound the hardy sealers praise, a wild and cheerful strain,
Who coast each creek and shore along or cross the billowy main;
Nor winter's storm nor seas alarm can daunt his daring mind,
Unknown to fear away he'll steer, old Neptune's place to find.

The merchant men cease voyaging when the wintry star appear,
The men-of-war are cruising far, propelled by southern air;
But hardy sons in Newfoundland wait not for season's change,
Mid ice and snow they'll daring go tho billows boldly rage.

The stormy month of March is come, boys, muster and prepare,
To leave your home o'er seas to roam and take the sealers fare;
Away, away, through storm and sea a trackless course to trace,
Where sea dogs lie and sea gulls fly, to find your favourite place.

When storms prevail to shorten sail aloft bold sealers go,
On slippery shrouds while piping loud the stiff northeaster's blow;
While slippery shrouds are piping loud, they mount the quivering yard,
The canvas roll, the billows foam, each wild lurch regard.

The gale increase take in close reef, our watchful skipper cries,
Luff, helmsman, to give relief whilst mountainous billows rise;
Let all but few descend below lest stormy seas o'erwhelm,
The smartest hands on deck must stand, the steadiest to the helm.

When night's dread noon, with cheerless gloom o'er shadow sea and land,
And breakers o'er the tempest roar proclaims the ice at hand;
With courage true our sealing crew for danger then prepare,
As they boldly wedge to the waters edge where none but sealers dare.

In yonder skirt begin the sport where many a white coat lies,
With gaff and gun each man has sprung to seize his distant prize;
Now, here and there, old seals appear, the white coats are at hand,
Your hauling gear and guns prepare and let the punts be manned.

Yon iceberg's wake has formed a lake that lees our course along,
To warp her through our sealer crew round bows and bulwarks strong;
Now the storm has passed we're stowed at last, our hatches battered down,
'Tis homeward now our barque shall plow, the sealers hopes to crown.

This song was written in 1842, years previous to the steamers going to the ice. Sailing vessels prosecuted the sealing industry in those years. Mr. Webber, of Harbour Grace, who edited a newspaper called The Conception Bay Man, *was the author. (Original note)*

Reprinted with permission from James Murphy, SONGS SUNG BY OLD TIME SEALERS OF MANY YEARS AGO. (St. John's: James Murphy, 1925), 12.

Song of the John Martin (1)

Come all ye jolly fishermen
Agoing to the ice,
Beware of the *John Martin*
And don't go in her twice,
For I was in her last spring
But I'll go in her no more,
If I cannot get a better berth
I'd rather stay ashore.
Chorus: — To my ring tooral laddy,
To my ring tooral lee.

Are you goin' to tap yer boots, Bill,
Yes, indeed, and that I am,
For tomorrow mornin' early
We must face the frozen jam,
So go and put the kettle on,
We'll have a cup of tea,
For the weather bow to-morrow
May be better than the lee.

Chorus: — To my ring tooral laddy
To my ring tooral lee.

The skipper he came forward
And he swore once or twice,
Took the kettle by the hangers
And he threw it on the ice.
Oh! I never felt so sorry
Since the day that I was born
When I saw my little "piper"
And it floating down astern.

Chorus: — To my ring tooral laddy
To my ring tooral lee.

Good luck to skipper "Nick" Ash,
For he's a clever man,
Sure he spied the laddios
Twelve miles in on the jam,
And following his council
We struck a heavy patch,
And it wasn't very long
When we filled her to the hatch.

Chorus: — To my ring tooral laddy,
To my ring tooral lee.

When we got in to the patch
The swiles were very thick
And the skipper he came forward
With a junk of a stick
Sayin' ye burnt all my lassy
And ye roasted all my pork,
So now ye mortal "sons of guns"
I'm goin' to make ye work.

Chorus: — To my ring tooral laddy,
To my ring tooral lee.

This vessel belonged to Carbonear, she was lost on her way from the Labrador to Carbonear, the fall of 1859. The name "laddios is what the writer called the seals, which is mentioned in the second verse. The song was written in the second verse. The song was written in the early forties or thereabout.
(Original note)

Reprinted with permission from James Murphy, SONGS SUNG BY OLD TIME SEALERS OF MANY YEARS AGO. (St. John's: James Murphy, 1925), 6.

The John Martin (2)

Come all ye jolly fishermen agoing to the ice,
Oh, beware of the *John Martin* and don't go in her twice.
For I was in her last spring and I'll go in her no more.
If I cannot get a better berth, I'd rather stay ashore.

With my ring to re la ring to lah rady oh.

Good luck to skipper Nick Ash for he is a clever man.
Sure he spied the "Laddioes" ten miles in on the jam
And following his council, we struck a heavy patch.
And it wasn't very long before we filled her to the hatch.

He went up on the topsail yard and he came down again,
And he looked in the binnacle to see how was the wind,
And he says unto the skipper, for you I've got good news.
Steer your course due north-north-east and you'll strike the "Laddioes."

It was down in the forecastle that a hell of a row arose,
It was all about the boiling of a kettle full of brewse.
The skipper he came forward and he swore once or twice,
He took the kettle by the hangers and he threw it on the ice;
I never felt so scalded since the day that I was born,
When I saw my little piper and it floating off astern.

Are you going to tap your boots, Bill? Yes, and that I am,
For to-morrow morning early we must face the frozen jam,
So go and put the kettle on and we'll have a bowl of tea,
For the weather bow to-morrow will be better than the lee.

Now when we got into the jam the swoiles were very thick,
And the skipper he came forward with a junk of a stick,
He said you burned all my lasses and you roasted all my pork,
So now you mortal sons o'guns I'm going to make you work.

Air: — "As I roved out one morning."

The following verses were composed in 1845 by Stephen Reardon of Perry's Cove, and they were very popular. The John Martin *was a brig owned and commanded by John Bransfield. The "Laddioes" was a nickname given to the seals.*
(Original note)

Reprinted with permission from Gerald S. Doyle, OLD TIME SONGS AND POETRY OF NEWFOUNDLAND. (St. John's: Gerald S. Doyle Ltd., 1927), 25.

Ferryland Sealer

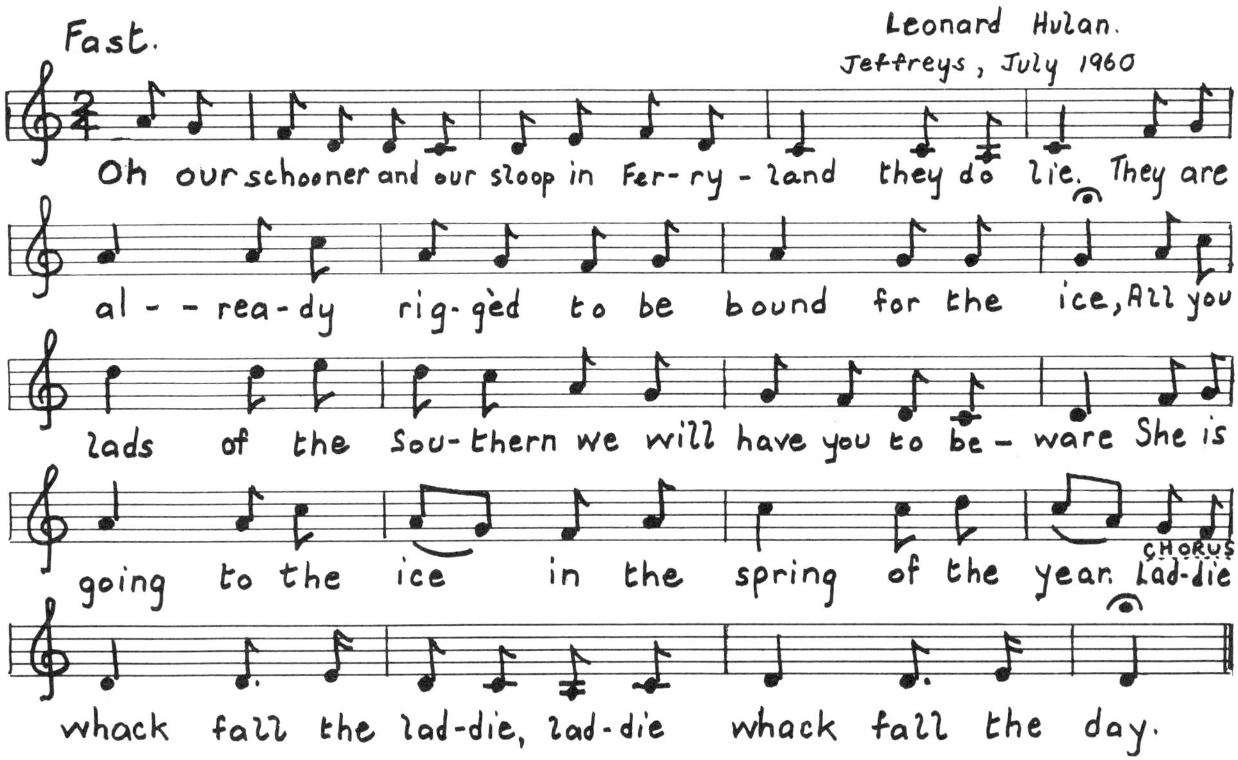

Oh our schooner and our sloop in Ferryland they do lie,
They are already rigged to be bound for the ice,
All you lads of the *Southern* we will have you to beware
She is going to the ice in the spring of the year.
Laddie whack fall the laddie, laddie whack fall the day.

We had vittles for to last more than two months at the least,
And plenty of good rum, boys, stowed away in our chest,
We will give her a rally for to praise all our fancy,
All our seals will be collected by the *William* and the *Nancy*.

Our course be east-north-east for two days and two nights,
Our captain he cried out, "Boys, look ahead for the ice!"
And we hove her about standing in for the land,
And 'twas in a few hours we were firm in the jam.

Oh our captain he cried out, "Come on, boys, and bear a hand!"
Our cook he gets the breakfast and each man takes a dram,
With their bats in their hands it was early to go,
Every man showed his action 'thout the missing of a blow.

Some were killing, some were scalping, some were hauling on board,
And some more they were firing and a-missing of their loads.
In the dusk of the evening all hands in from the cold,
And we counted nine hundred fine scalps in the hold.

Oh now we are loaded and our schooner she is sound,
And the ice it is open and to Ferryland we're bound,
We all gave her a rally for to praise all our fancy,
Our seals they were collected by the *William* and the *Nancy*.

We are now off Cape Spear and in sight of Cape Broyle,
We will dance, sing, carouse, my boys, in just a little while,
We will soon enjoy the charms of our sweethearts and friends
For it will not be long before we're down to the bend.

A harp and whitecoat sunning on the ice. *Photo by Hal Andrews.*

The Land of Fish and Seals

Let Sunny India her wealth proclaim, her gorgeous glowing sky,
Her silken stores, her golden veins, and flowers of every dye;
We envy not her gaudy show where death insidious steals,
For wealth's bright diamond deck our brow in the land of fish and seals.

Let Italy and France and Spain their vine-clad valley's praise,
Let Greece and Rome take up the strain and sing of bye-gone days;
Of classic fames and gorgeous names which fame's loud trump reveals,
We boast not of past glory in the land of fish and seals.

No great immortal names are ours whose deeds our annals trace,
No sunny skies, no golden mines, nor vines our valleys grace;
And yet your ice-girt Isle can claim the true, the brave, the leal,
For slave and tyrant we disdain in the land of fish and seals.

We boast not of our mighty dead, but of our living brave,
Who the frozen ocean fearless tread, or skim the angry wave;
Heaven's blessing on their gallant hearts our country's pride and weal,
Who brave the perils of the deep in quest of fish and seals.

May plenty cheer our hardy sons and fortune on them smile,
If milk and honey are not ours may we have lots of oil;
It will make each face shine with joy, each bosom grateful feels,
And give contentment, joy and peace in the land of fish and seals.

This song was written by the late Mrs. Peace, a Scotch lady, but a resident of St. John's in the fifties. (Original note)

Reprinted with permission from James Murphy, SONGS SUNG BY OLD TIME SEALERS OF MANY YEARS AGO. (St. John's: James Murphy, 1925), 19.

Driven into Spaniard's Bay

'Twas on the very first day of March,
To the ice, boys, we were bound;
The wind came up from east northeast,
And we had to heave her round.

We had like to run her on Green Head,
Before we saw the land;
And a block fell from the masthead
Which had like to kill a man.

The skipper and another man,
To the topmast yard they flew;
There's ice and shoal to leeward, boys,
And we'll bear on them, too.

The wind came up with a terrible roar,
With snow we could not see;
And when we found out where we were
We were up in Conception Bay.

We ran her up to Spaniard's Bay,
Not knowing what to do;
We hove two anchors from her bows
And then we brought her to.

It being a dark and stormy night,
We stayed till break of day;
And on the seventeenth of March
We sailed from Spaniard's Bay.

Come Down with the Killock

Come down with the killock
And out with the line;
Of fish about here, boys,
There is a good sign.

'Tis early in April
What matter for that;
We are not like the fools
Who are hunting for fat.

Hunting for fat, boys,
They never may get;
And losing the fish
They could get in their net.

A bird in the hand
Is worth two in the bush;
And off to the ice
Go fools with a rush.

Terra Nova Seal Fishing

Ye talk o' this, and talk o' that,
An' lots o' ither — silly chat;
But list tae me — I ken ye weel
Wad like tae hear aboot the seal.

Eh me! What wonders here we see
They'd gar ye laugh wi' greatest glee;
But as ye canna see the sport,
Tae a faint description I'll resort.

In winter, when the snow does fa',
Aff gang the sealers far awa'
In ships, wi' crews a hunder men,
Resolved tae cause what death they can.

Talkin' o' death brings me in mind
O' a toast which seemed tae be unkind,
But an explanation makes amends,
'Twas only "Death to our best Friends."

But noo, I'm rather affi my story,
Wi' bein' in ower big a hurry
Sae here, I'll jist begin again,
An' try tae to keep the proper strain.

The ships gang richt tae the frozen sea,
For there the seals are sure tae be;
Sae no, I'll tell ye if ye're willin'
Tae listen tae the style o' killin'.

Sae money men hae sticks ca'd bats,
Wi' which they fell the youngish brats;
Ithers hae guns, wi' guid wide bores,
An shoot the ould ones by the scores.

The young, ye see, we manage fine —
The differ frae the aulder kin' —
As ae stroke frae a batsman's pole
Puts them at once into control.

Yes! but they're sometimes very cunnin',
For after gettin' quite a stunnin',
An look as dead, revive again,
Try tae escape, but a' in vain.

The auld ones mak' an awfu' fuss,
An' great resistence offer us;
They stan' on end almost upright,
An' gang ower near we'd feel their bite.

When killed they're laid upon the back,
Cut up the middle in a crack,
The intralls then are thrown away,
The rest's for use some future day.

The fat when taken from the skin,
Is inches deep, and that's no thin;
By steam machines this fat we boil —
The substance made a ca'd seal oil.

The skins we noo wi' salt do serve,
The fur an' beauty tae preserve;
An' if this thing wasna dune,
Ev'ey hair wa'd leave the skin.

The seals hae furs o' diff'rent styles,
An' hereabouts they ca' them swiles,
But that's nae odd — ye ken, at least,
That its a very usefu' beast.

The Harp seal skins has pretty spots
Of black, along wi' fainten dotes.
Shining like gold, sae bright an' clear,
An' worth as much, tae people here.

The Hood is quite o' anither grade;
Her skin is just a light and shade,
But looks rale weel, an' tae be short,
The Hoods I think a pretty sort.

The Bedlamer seal is o' either class,
But for neither auld or young must pass,
She's valued as a middle aged dame,
Ye see there's somethin' in a name.

Then, agaín, the whelps are ca'd whitecoats,
Their furs are o' a yellow white,
An' really are a charmin' sight.

But noo', methinks, I'll stop my whistle,
It's made a very puir epistle,
Yet hope I've made ye understand
About the seals o' Newfoundland.

Robert Brown
Newfoundland, June 10, 1867

Reprinted from THE HARBOUR GRACE STANDARD.
Harbour Grace, August 7, 1867.

A Newfoundland Hero

The fierce wind moaned among the cliffs of rugged Labrador,
The wild waves dashed with thund'rous sound against the rock-bound shore;
The snow that dimm'd the noon-day sun fell on the muffled form
Of one who, blest with manhood's strength, defied the raging storm.

With step that spoke a fearless heart, a strong and steadfast will,
He strode along the rocky beach and crossed the barren hill;
Nor wind nor sea nor blinding snow he heeded as he passed,
But by some secret force impell'd he hurried thro' the blast.

Ah, hark! was that a cry for help borne on the angry breeze?
It rose above the tempest's din, above the raging seas;
It came from o'er the waste of foam, that chill, despairing cry,
The sad appeal to Heav'n addressed of men about to die.

It reached the throne of Him who'd need but will to be obeyed,
Who loves to hear weak human tongues invoke His puissant aid.
'Twas by His great omniscience the hero of this tale
Was led to seek that fatal spot in such a fearful gale.

As down the craggy path he came, the cry fell on his ear,
And well he knew it had its source in dreadful shipwreck near.
He gazed far o'er the breaking surf, into the snowy air;
No wonder that his eye grew dim at what he saw out there.

Some hundred fathoms from the shore, upon a reef of rock,
A bark had struck, while spars and keel were shivered by the shock.
The jagged point on which she lay had pierced from keel to deck,
And pale with fear the trembling crew were clinging to the wreck.

E'er and anon the crested waves upon her rushed amain,
As if they in their mad career would wrench her planks in twain.
A moment JACKMAN gazed upon this scene of fearful woe,
Then flung his boots and coat apace upon the drifting snow.

His lips were set in firm resolve, as down the slope he dashed,
Ne'er wav'ring tho' the surging waves around him roared and crashed,
And plunging in the yeasty mass, he sank and rose again,
Then boldly struck out for the reef to save the drowning men.

On, on he swam, while in his face the cold, fierce sea-breeze blew,
And o'er his head in show'rs of spray the briny waters flew.
Tho' many a time a billow huge above him raised its crest,
Despite its force he pierced it through and swiftly onward pressed.

And when his hand the bulwark grasped, not long he rested there,
But space to raise his voice to God in short, tho' fervent prayer;
Then seized he him who nearest stood, and 'mid the noise and strife,
"Come on," he cried, "hold fast to me, and I will save your life."

Then shoreward, bravely on he swam, the heaving waves among,
While to his waist, with voice-like grasp, the shipwrecked sailor clung.
Despite the furious element, they safely reached the land,
And with their garments dripping brine, stood on the sea-beat strand.

Then JACKMAN cried, "Haste! haste, my friend, across yon sandy fell,
And to the men you'll find o'er there, these woeful tidings tell.
Stay! bid them bring with all their speed life-buoys and ropes to me,
And by God's help we yet will cheat this all-devouring sea."

So spake he — and as from his side the rescued man rushed off,
He turned and darted once again into the seething trough.
From shore to wreck, from wreck to shore, ten times the seas he braved,
And ere the men with ropes arrived, ten more lives had been saved.

"Now, quick, my man," our hero cried, "bring here that buoy and rope,
And let your knots be such as wild and raging seas won't ope."
Then hurriedly the trusty buoy he 'neath his shoulders drew,
And nerved his tired and aching limbs for what he'd still to do.

Two ropes were fastened to the buoy, one round him coiled he bore,
The other lay in friendly hands upon the snow-clad shore.
Into the surf he leaped once more, while many an earnest prayer
From those on land and on the wreck rose thro' the quivering air.

And oft they saw his form submerged, and thought he needs must fail;
But still he reached more he came and went across that

And oft they saw his form submerged, and thought he needs must fail;
But still he reached and tied the rope around the vessel's rail.
Sixteen times more he came and went across that foamy tide;
Each time a grateful human heart throbbed wildly at his side.

At last upon the beach he stood, with shiv'ring limbs and weak;
The rescued men, in silence, looked the thanks they could not speak.
"All have been saved, thank God," he cried; "come, let us o'er the hill,
For fire and clothing both need we to counteract this chill."

"Who spoke? What would you now, my man?" he
queried in surprise.
"Oh, sir! a weakly woman still in yonder cabin lies."
"What! swim out to the wreck again? — 'tis madness,
sir, I say;
She's dead — if not, she is so weak, she'd perish on the
way."

"Alive or dead, she'll not stay there upon the lonely
sea,
And if she dies the death to-day, the fault won't rest
with me."
Tho' hard they pressed him not to go, their efforts were
in vain,
In haste he donn'd that well-tried buoy, and dared the
waves again.

The deck was reached, the cabin door he shivered with
a blow —
Then bore the pale and fainting form from out the
berth below.
The fierce and angry ocean strove its utmost to entomb
Our hero, and his helpless charge, within its boundless
womb;
But, drawn by strong and willing hands, at last he
reached the shore,
Where cheers arose from grateful hearts, that drowned
the tempest's roar.

The hero of the poem was the late, Capt. William Jackman, a noted sealing captain of his day.
(Original note)

Reprinted with permission from James Murphy, SONGS OF OUR LAND: OLD HOME WEEK SOUVENIR. (St. John's: Telegram Print, 1904), 44.

Captain Bill Ryan Left Terry Behind

Air: Paddle your own canoe.

Terry is a fine young man,
But he has a lot of "chaw,"
He thought to do the devil and all,
When he got the *Esquimaux*.

The *Mary Joyce* is stuck in the ice,
And so is the *Osprey* too.
Captain Bill Ryan left Terry behind,
To paddle his own canoe.

Though Terry is a fine young man,
He couldn't cut no shine,
Although he had the *Esquimaux*,
And Dawe has got the *Lion*.

The *Wolf* and the *Lion* kept very
good time.
And so did the *Panther* too.
But Captain Billy Ryan left Terry
behind
To paddle his own canoe.

This was probably one of the best known sealers song in its day. Capt. Bill Ryan and Terry, and the other names mentioned, were real characters and the incident told in the verses was an actual occurrence.

Reprinted with permission from Gerald S. Doyle, OLD TIME SONGS AND POETRY OF NEWFOUNDLAND. (St. John's: Gerald S. Doyle Ltd., 1927), 65.

Sealers' Song

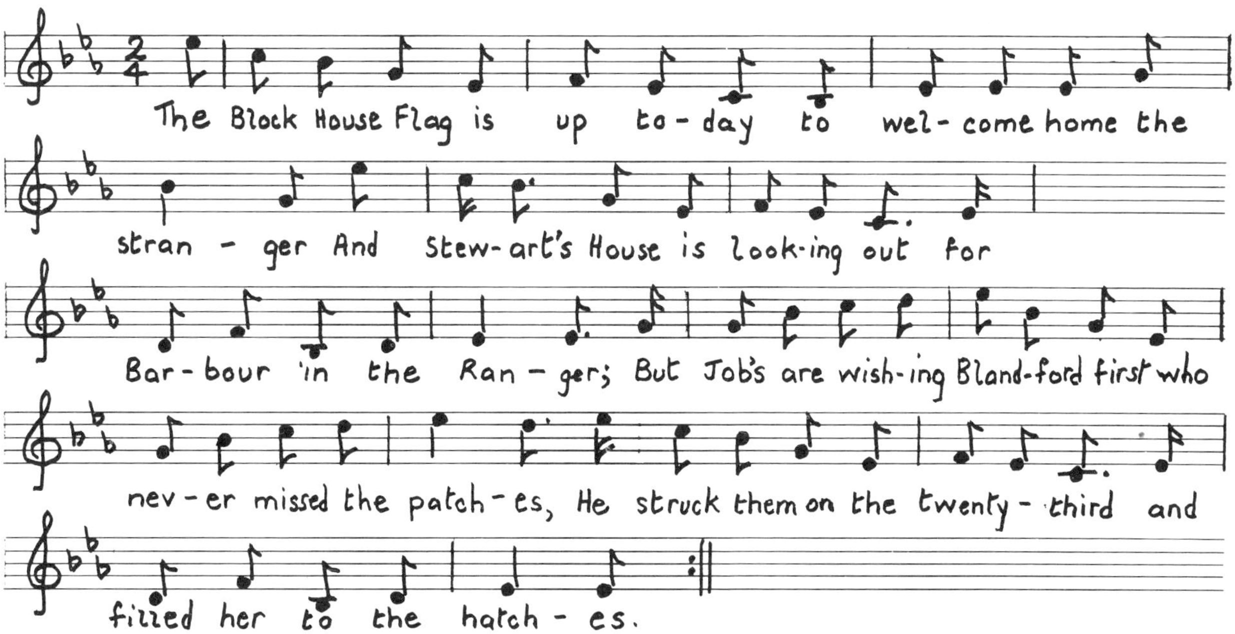

And Bowring too will bet a few
On Jackman in the *Howler*,
The little *Kite* she bore in sight
With Billy Knee the Jowler.

The first of the Fleet is off Torbay,
All with their colours flying:
And Girls are busy starching shirts
And pans of beefsteaks frying.

We left you see with Billy Knee
Bound Home with colours flying,
But were forced to stay at Trinity Bay
Two weeks or more there lying.

Though short of grog still lots of prog
To bring us home quite hearty
Each Trinity Dove fell wild in love
With Walsh and Luke McCarthy.

Oh in the spring the flippers bring
To lawyers, clerks or beagle;
We fought brave Neptune up and down
And carried home the *Eagle*.

Though some may sing of lords or kings,
Brave heroes in each battle,
Our boys for fat, would gaff and bat,
And make the whitecoats rattle.

They's kill their foe at every blow
(Was Waterloo more grander?)
To face, who could, an old dog hood
Like a plucky Newfoundlander?

We danced on shore in Bremmner's store
The darling girls were dancers;
Jemina Snooks our Boys would hook
At every set of lancers.

I felt afraid of the fuss they made
Of each confounded villain;
I thought the floor would leave the store
At the Trinity Bay Cotillion!

Don't talk to me of balls or sprees,
You never saw such a party
That time on shore at Bremmner's store
Made all feel good and hearty.

For at a dance no girls can prance,
Nor dress in style more grander
For an irish reel, that takes the heel,
To please a Newfoundlander.

So here's success to Susie Bess,
And girls from all outharbours,
For a kiss set in on a sealer's chin,
Which never saw the barber.

The Sealing Trip of the S.S. Greenland 1891

All ye who love old Newfoundland
And her Sons who plough the sea
Midsummers suns and winters' snows
So bold and fearlessly
Please pay attention for a while
And I will sing to you
A song about the *Greenland*
And her hardy sealing crew.

We're just arrived at Harbour Grace
From the Southern white-coat patch,
The hold with oily pelts is stowed
Chock up to every hatch.
We were the first to strike them
A fact, deny who can
And not a ball was chewed for us
By our sister ship the *Van*.

When the Captain left the bridge to-day
And on the wharf did stand,
Full many were the friends he met
And shook his greasy hand.
He's a hardy Newfoundlander
Henry Dawe it is his name,
And he is a bold commander
When on the raging main.

On the morning of the tenth of March
From the Offing of Greenspon'
To get her "links in motion"
It did not take us long.
And Freddy was so anxious
To see that all was right,
He went "round her" like a paper man
'Till twelve o'clock that night.

Our pumps were in good order
The condenser it was clean,
The pistons worked like magic
And were well supplied with steam.
We were all in good condition
To face the northern jam,
And to keep the head position
The chief "rove" every man.

The Captain wore a smile that day,
Upon his handsome face,
And he says "Those Dundee Grayhounds boys
For us will have a chase.
So loose your topsails, fore and aft,
To catch the Sou' West wind,
And down off the Grey Islands
The seals we'll surely find."

A family of hoods was first
On us to make a call,
And on the evening of the twelfth,
We heard a white coat bawl.
Then next morning very early
Full equipped was every man,
With spirits right, soon as 'twas light
The seals commenced to pan.

The slaughter then was dreadful
'Tis useless to describe,
From East to West for miles around
The Ice was crimson dyed.
Sharp knives and bats did deadly work
And when the day was done
Twice seven thousand scalps were flagged
Beneath the setting sun.

And, mind me, Captain Harry
Never swore upon his crew,
He knew their sterling value
And their duty they would do.
And when the ship was loaded,
And we were homeward bound,
At the calling of the roll, each man
Turned up both safe and sound.

And as we neared the Harbour Bar
We steamed her slowly in,
With her colors gaily flying
And as deep as she could swim.
And people gathered on the wharf,
From every street and lane,
To welcome back the Sealers
From off the stormy main.

And now we're home for Easter
At the "hop" we'll swing the girls
With their neat wire form improvers,
And Dolly Varden curls.
"Nice" folks may perhaps laugh at
But they don't understand,
That they boys in oily jumpers,
Are the pride of Newfoundland.

And now three cheers for Captain Dawe
And long may he command,
A gallant ship and hardy crew,
From dear old Newfoundland.
And I hope success he'll always get
When in the Northern seas,
And may his "big jib" always draw
Filled with a moderate breeze.

Composed by one of her crew.
(Original note)

Reprinted from THE HARBOUR GRACE STANDARD.
Harbour Grace, 1891.

Trinity Bay Tragedy

Ye hardy sons of Newfoundland, that tread life's rugged way,
That know the key of many years, that now have passed away,
Draw near while I relate you an awful tragedy
That did befall our hardy sons down north at Trinity.

In eighteen hundred ninety-two, on February twenty-seven,
The morning broke with brilliant sky, that brightly shone the heavens,
The seas were smooth and tranquil, all nature seemed at rest;
In search of seals our boat soon sped upon the billows' crest.

That morning when we left our homes and lanched (launched) out from the land,
We little apprehended what dangers were at hand
We bade good-bye to those we loved, as we ofttimes done before,
Not thinking in our eager chase we'd see those friends no more.

Now same for to go pleasantly as we skimmed o'er the bay,
Till bellows swole in frenzy wild and struck us with dismay.
Our small boats reeled to wind and sleet, whilst each man plied his oar,
With eager hopes at every stroke for to regain the shore.

Yet some they had succeeded in reaching of the land.
The sight was most appalling to see that icy hand,
Combined with frozen icicles, with gusts of wind and snow,
That fell upon those stalwart limbs and laid their victims low.

What awful sound is this we hear, comes floating o'er the lea?
It is no sound of merriment; it is no revelry!
It's borne upon the northern blast across the stormy sea;
It is the cry for human help; it comes from Trinity.

But, alas, that help is not at hand and they are doomed to die
That awful death by freezing on cold ice-fields to lie;
No human thought can picture the anguish that they felt
Not till the vital spark had fled and life became extinct.

Next morning dawned with ghastly form and frowningly they looked down
On that awful ravages with corpses strewn around,
But yet there's some with frozen limbs that struggled through the night,
For their life sake one effort make and landed at Heart's Delight.

They left to tell the doleful tale, while kind friends gathered round,
And many a pitying glance was cast and silent tears flowed down;
One of those being the doctor's son so quickly called away
In the ocean deep in the long last sleep in the unconscious sea.

The mother cried in frenzy wild, her widowed hopes are o'er,
The father of her smiling babe on earth she'll see no more;
Bold death with its untimely grasp has taken them away,
Until the sea shall yield its dead on that great judgment day.

And may the God of mercy that died the bold to save
Extend His richest blessings to each mother and her babe,
And may they bow submissively unto His holy will,
For He is the great Omnipotent that bids the sea be still.

Died on the Ice Fields

The white, rugged ice-floe came gliding along
On the ocean, with seals scattered o'er,
And the eyes of the fisher-folk sparkled with joy
As it pressed closer still to the shore.

There was sharp'ning of sheath knives and seizing of "bats,"
There was gathering of "tow-lines" in haste;
And the heart of the youth in his teens swelled with pride
As he buckled a belt round his waist.

With heart light as air they pushed out o'er the ice,
There their "bats" raised a torrent of blows
On the harps, while the keen, glittering sheath knives soon gave
Them the "sculp" they required for their "tows."

In the midst of the bustle the fickle wind changed,
And the ice 'gan to move from the land;
There was slipping of "tows," there was running for life,
'Mong the men of that brave little band.

Richard Parsons, of Flat Rock, and his eldest son
(The younger had gone home before)
Were speeding along when the weary boy cried;
"I am tired; I can travel no more."

His garments, brine drenched, were now stiff with the frost,
His limbs had grown helpless and numb;
And the father with anguish untold was o'whelm'd
Lest the boy should e'er morning succumb.

His own clothing straightway he tore from his back,
And in it did quickly enfold
The shivering form of his perishing boy,
In an effort to keep out the cold.

Then clasping him close in a frantic embrace,
Thro' that dark night he did strive,
By breathing his own warm breath o'er his face,
To keep his chilled off-spring alive.

He reck'ed not of hunger, of cold or fatigue,
With his son on the brink of the grave;
And when found in the morning his pulse was as weak
As the boy he had labored to save.

Kind hands bore them shoreward, and still on the way
Another sad sight their eyes met,
The form of the younger boy, thought to be safe,
On the ice lying silent in death.

Vain, vain, is the effort in words to portray
The depths of that fond parent's love;
It can only be judged and rewarded one day
By an all-seeing Father above.

Tribute to the heroism of Richard Parsons, of Flat Rock.

The song was written by P.J. Dyer, a native of St. John's, but now a resident of the United States. The... incident, in which a son of Richard Parsons, of Flat Rock, near St. John's perished on the ice floe, took place on March 24th, 1894.
(Original note)

Reprinted with permission from James Murphy, SONGS SUNG BY OLD TIME SEALERS OF MANY YEARS AGO. (St. John's: James Murphy, 1925), 18.

Loss of the Maggie

Ye fishermen who know so well
The dangers of the deep,
Come listen to a dreadful tale
And join your tears to weep
For the loss of the schooner *Maggie*
And thirteen precious lives
Which leaves so many homes bereft
Of husbands, sons and wives.

At ten forenoon, November five,
The *Maggie* sailed away
From happy homes near Brooklyn,
In Bonavista Bay.
Light winds did waft her on her course,
Light hearted was her crew,
When the shades of e've were falling fast
The city come in view.

Our hopes ran high, our hearts were glad,
We soon would tread the shore,
And turn to cash the fruit of toils
Upon the Labrador.
The city lights did seem to greet
And welcome us to town,
When Captain Blundon cried, "my boys"
There's a steamer bearing down.

Like unrelenting monster fierce
That seeks its prey to get,
She bore straight on us, but we hoped,
Her course would alter yet.
We shouted loud in wild despair,
Too late, an awful crash;
Next moment o'er our shattered craft
The hungry waves did dash.

The scene that followed then, oh! God,
'Tis branded on my brain,
And rather would I join the drowned
Than witness it again
When shrieks heart-rending pierced the air,
In a desperate fight for life,
As brother saw a brother sink
And husband saw a wife.

Of twenty-three who left their homes
Upon that fatal morn,
Thirteen, alas, are hushed in death
And never can return.
And the name of the steamer *Tiber*
Will tell in days to come
On the minds of Brooklyn people
Like the sound of funeral drum.

Note: — The schooner Maggie, Blundon master was lost Nov. 5 1896. She was cut down in the narrows of St. John's by the S.S. Tiber on a passage from Brooklyn, B.B.; 13 out of 23 souls on board were lost.
(Original note)

Reprinted with permission from James Murphy, SONGS AND BALLADS OF NEWFOUNDLAND, ANCIENT AND MODERN. (St. John's: James Murphy, 1902), 61.

The Wreck of the Maggie

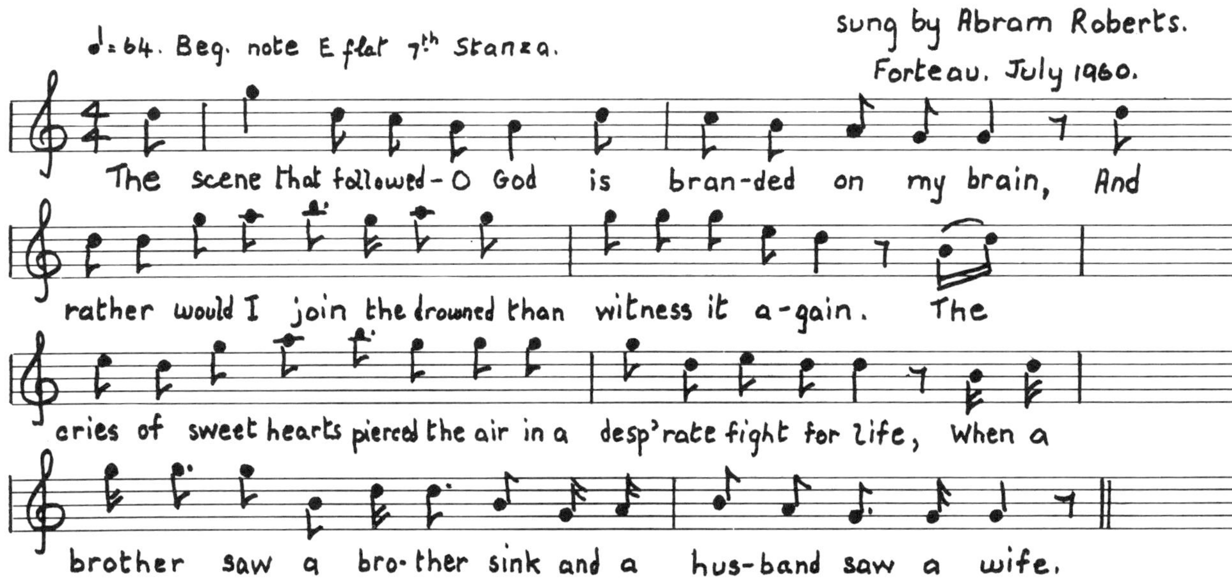

At ten forenoon November five the *Maggie* sailed away
From her happy home near Brooklyn in Bonavista Bay.

Light winds did waft her on her course; light hearted was her crew,
And Friday evening off the Narrows the city came in view.

Our hopes ran high; our hearts were glad; we soon would tred the shore,
And turn to catch the fruits of toil upon the Labrador.

The city lights did seem to greet and welcome us to town,
When Captain Blunden cried, "My boys, there's a steamer bearing down."

Like a raging monster fierce that seeks her prey to get,
She bore straight on us and we hoped her course would alter yet.

We shouted aloud in a wild despair — too late — an awful crash!
Next moment o'er our shattered craft the hungry waves did dash.

The scene that followed (then) — O God — is branded on my brain,
And rather would I join the drowned than witness it again.

The cries of sweethearts pierced the air in a desp'rate fight for life,
When a brother saw a brother sink and a husband saw a wife.

And the name of the steamer *Tiber* will fall in days to come
On the ears of the Brooklyn people like a sound of a funeral drum.

Mr. Roberts, now 84, has known this song all his life. He learned it 'from a sailor.' It refers to the wreck of the schooner Maggie making into St. John's with a catch of seals and being run down by the steamer Tiber.
(Original note)

The Spring of '97

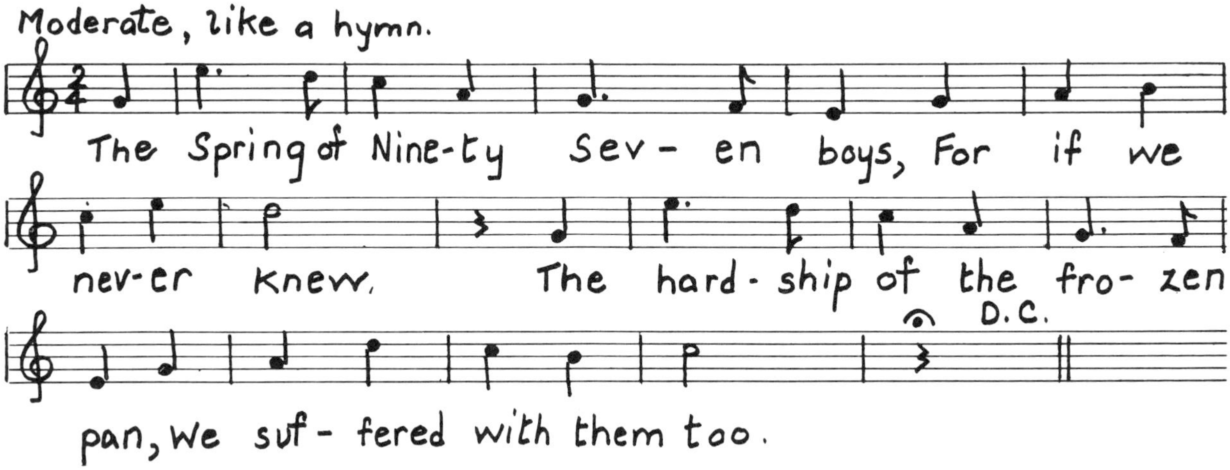

The Spring of Ninety Seven boys,
For if we never knew
The hardship of the frozen pan,
We suffered with them too.

We struck the seals off Cabot Isle,
Five days out from port;
We thought to have no long delay,
And loading would be sport.

Eleven thousand prime young harps,
We put on board that day;
And sixteen thousand more that night,
We safely stowed away.

Next day a storm broke off the ice,
Which ripped through our port bow;
To save a life it was enough,
For each man to look how.

We suffered awful hardship then,
When often cold and wet;
And all the dangers of that Spring,
We never will forget.

'Twas bad enough till Death it came,
Made our condition worse;
For one of our brave Harbour men,
Died on the twenty-first.

When doing duty like a man,
Our comrade stopped to fall;
From which he suffered much too dear,
'Till Death did give a call.

Then ship-mate John Sparkerell,
Such hardship could not stand;
And on the twenty-ninth he sighed,
Passed on to a better land.

'Tis on the second of April,
Poor Swain did pass away;
Soon followed poor Mike Sullivan,
Belong to Caplin Bay.

He bravely fought to keep up fit,
To see his friends once more;
But never will his footsteps tread,
Upon the Southern Shore.

Oh Newfoundland, in north and south,
Around in every bay,
Take pity on our poor countrymen,
Caught up in this sad way.

They bravely died in duty's path,
And God will mercy shed;
Those men who strived so hard on earth,
To earn a crust of bread.

The Greenland Disaster (1)

They sailed from St. John's Harbour all on the tenth of March,
Commanded by Captain Barbour, the ice-fields for to search.

Listen to my mournful story; your grief it will renew,
When I relate the hardships that befell the *Greenland's* crew.

Her course it lay into the North; she boldly sailed away —
Down to the North, passed by the Funks, she still kept on her way.

No danger seemed to threaten her, this gay and gallant boat,
And on the twelfth, I heard them say, they took their first whitecoat.

From then until the twenty-first, all seemed so bright and gay,
And for to get a paying trip they killed and panned away.

It crowned their labours with delight, the prospect being so great;
They did not know the grief and woe that on them did await.

'Twas early on the twenty-first, just by the morning light —
The Captain gave his orders, all with a cheerful voice.

"All hands, all hands upon the ice! Be ready, one and all!"
And each man then most cheerfully reponded to his call.

But in a short time after that the storm it did arise,
Boreas blew with vengeance, which darkened o'er the skies.

Those poor, unhappy creatures, they knew not where to go;
They could not find protection from the bitter frost and snow.

They then drew close together, their freezing limbs to warm;
It was a small protection from that wild and bitter storm.

They raised a prayer most fervently to Him above the sky;
They cast one mournful glance, and then laid down to die.

And then the *Greenland* came in view — oh, what a dreadful sight!
Twenty-five stiff, frozen corpses lay dead upon the ice.

Those twenty-five were brought to land, but shocking to relate,
There were twenty-three still missing — made the number forty-eight.

And now they fill a watery grave — from home and friends, away,
Until the death roll shall be called upon the Judgment Day.

"May the Lord have mercy on their souls," will be our fervent prayer,
And may they rest in heaven, always free from earthly care.

— Traditional

The Greenland Disaster (2)

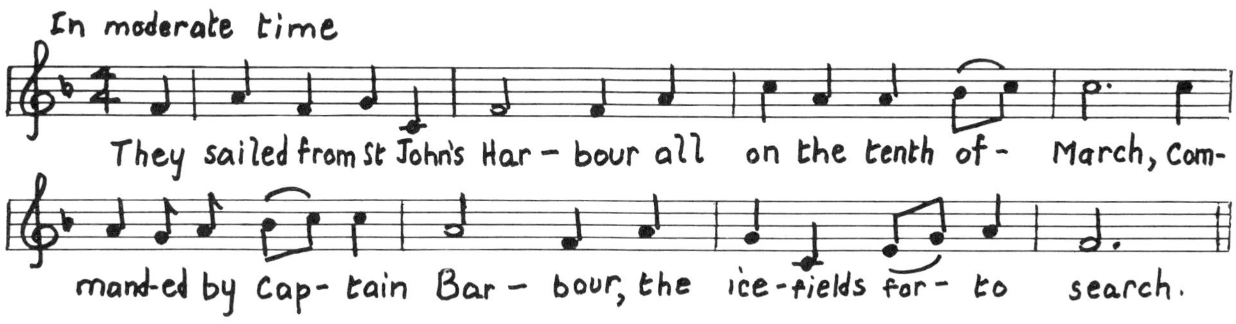

Ye tender hearted Christians, I hope you will attend
To these few feeling verses that I have lately penned.
Listen to my mournful story; your grief it will renew,
When I relate the hardships that befell the *Greenland's* crew.

They sailed from St. John's Harbour all on the tenth of March,
Commanded by Captain Barbour, the ice fields for to search;
With colors flying gaily they gave three hearty cheers,
But mark what followed after, you quickly shall hear.

Her course it lay into the North, she boldly sailed away
Down to the North, passed by the Funks, she still kept on her way;
No danger seemed to threaten this gay and gallant boat
And on the twelfth, I heard them say, they took their first white coat.

From that until the twenty-first all seemed bright and gay,
And for to get a saving trip they killed and panned away;
It crowned their labors with delight the prospect being so great,
They did not know the grief and woe that on them did await.

It was early on the twenty-first just by the morning light
The Captain gave orders, all with a cheerful voice:
"All hands, all hands upon the ice, be ready one and all!"
And each man then most cheerfully responded to that call.

But in a short time after that the storm king did arise,
Boreas blew with vengeance which darkened o'er the skies.
Those poor unhappy creatures, they knew not where to go;
They could find no protection from the bitter frost and snow.

They then drew close together their freezing
limbs to warm;
It was a small protection from that wild and
bitter storm.
They raised a prayer most fervently to Him
above the sky;
They cast one mournful glance all round and
they laid down to die.

When the *Greenland* came in view, O, what
a dreadful sight!
Twenty-five stiff frozen corpses lay dead
upon the ice.
Those twenty-five were brought to land,
but, shocking to relate,
There are twenty-three still missing, which
number forty-eight.

This is one among the missing, from
St. Brenden he came;
He was an honest fishermen,
Mike Hennessey by name.
On Tuesday night they laid him down upon
the ice to sleep;
Boreas blew a bitter squall which threw him
in the deep.

And now he fills a watery grave from home
and friends away,
Until the death roll shall be called upon the
Judgment Day.
"May the Lord have mercy on his soul,"
will be our fervent prayer,
And may he rest in heaven, free from all
earthly care.

One William Heaton from Harbour Grace,
that promising young man,
His parents' joy and hearts' delight, describe
their grief who can —
To see their dear and darling boy cut down
in his bloom;
With heavy sighs and mournful cries they
laid him in his tomb.

But now he is gone, that gallant boy, and
why should they repine?
There is many a one as well as him, have
left they friends behind;
There are mothers, wives and orphans who
are left for to complain
For those who in the *Greenland* sailed and
never returned again.

Now to conclude and finish, I have a few more
words to say.
I hope you all will join me to the Lord to pray
To comfort those that are left behind and to them
peace restore.
From grief and woe and broken hearts for friends
they'll see no more.

Reprinted with permission from Gerald S. Doyle, OLD TIME SONGS AND POETRY OF NEWFOUNDLAND. (St. John's: Gerald S. Doyle Ltd., 1940), 40.

The Greenland Disaster (3)

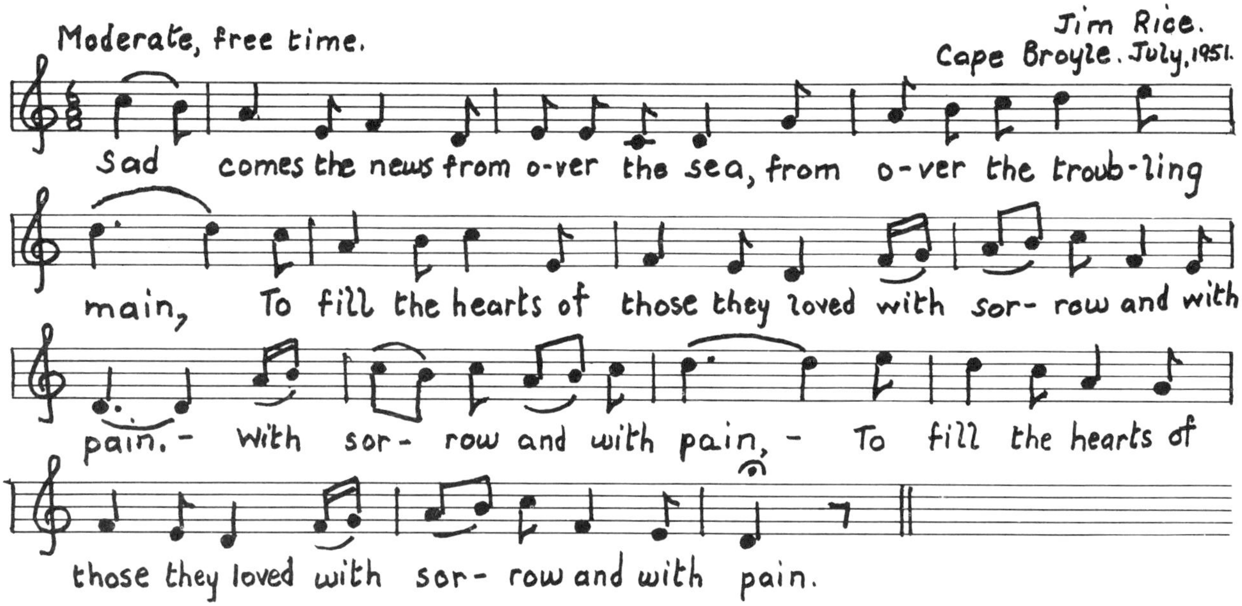

Sad comes the news from over the sea, from over the troubling main,
To fill the hearts of those they loved with sorrow and with pain.
With sorrow and with pain,
To fill the hearts of those they loved with sorrow and with pain.

Oh it's less than three short weeks today they left their native shore,
But all, alas, they never returned to see their friends no more.
To see their friends no more,
But all, alas, they never returned to see their friends no more.

The *Greenland* sailed the tenth of March, her crew in spirits gay
Stood brisk upon the barricade as she steamed out that day.
As she steamed out that day,
Stood brisk upon the barricade as she steamed out that day.

And with a crew of fine young men who left in spirits light,
Not thinking that they should pass off on that fearful cold March night.
On that fearful cold March night,
Not thinking that they should pass off on that fearful cold March night.

They struck the seals St. Patrick's day and then the work began,
With thirteen thousand hoist on board, seven thousand more did pan.
Seven thousand more did pan,
With thirteen thousand hoist on board, seven thousand more did pan.

On to the thirty-first of March, four watches less that day,
A blinding snow storm soon came on and they got cast away.
And they got cast away,
A blinding snow storm soon came on and they got cast away.

There's two young Quidi Vidi boys, there's Chillie and Court Down,
There's William Collins from Torbay, and two from St. John's town.
And two from St. John's town,
There's William Collins from Torbay, and two from St. John's town.

The others from the northern bays and six from Harbour Grace
Who on that wild and terrible night a fearful death did face.
A fearful death did face,
Who on that wild and terrible night a fearful death did face.

But may the great Almighty God their troubles soon make light,
And ease the broken hearts of those who lost their friends that night.
Who lost their friends that night,
And ease the broken hearts of those who lost their friends that night.

Sealers and ships preparing for the sealing voyage, c. 1900.

Courtesy of the Provincial Archives of Newfoundland and Labrador (PANL).

Written in Memory of the 48 Men Who Lost Their Lives in the S.S. Greenland Sealing Disaster of Monday, March 21st, 1898. (4)

A wail from the stricken mothers and wives,
Who mourn for the sealers brave,
Who weep for their loved and dear ones' lives,
Gone out on the frozen wave.

The air is filled with direful woe,
For the lost two score and ten;
Who perished afar amid the floe —
Our brother fishermen.

'Tis less than three short weeks ago
Since the *Greenland* left the land;
And hearts felt gay, for the coming fray,
In the breasts of the gallant band.

A golden sun in an azure sky,
Smiled down on the waves so blue;
And quietly kissed of the morning mist,
And cheered the hardy crew.

No dread of danger was in the air,
All hushed was the tempest's roar;
And the Boreas breeze seemed rocked at ease
Far north in the Labrador.

All hail for the blood-red harvest now,
The vessel has struck the patch.
"Out, kill and pan, work every man,
'Till the ship's filled to the hatch."

Right bravely wrought these gallant men,
Nor thought of approaching harm,
'Till the northern giant's sullen roar
Betokened the coming storm.

And then the scramble for ship and life
O'er icefields, wild and wide.
Oh! the story of those who suffered and lived,
And those who suffered and died!

Oh! mothers who live by the northern seas,
Oh! wives of the perished brave;
Our prayers with thine to-day combine,
And looking beyond the grave,

We feel that whether your dear one lie
Beneath the sea or the sod,
That he who falls where duty calls,
Will rest in peace with God.

Reprinted from THE HARBOUR GRACE STANDARD.
Harbour Grace, 1898. Written by M.A. Devine, March 28, 1898.

Greenland Disaster (5)

All pious hearted christians,
Please harken unto me,
And hear of Forty Eight poor men
Who lost their lives at sea.
Who leave in grief heart broken wives,
While Orphen Children cry,
And poor old Mothers bent with age
The loss of her dear boy.

The Greenland, sailed the 10. of March,
With all the sealing Fleet,
And Captain Barbour in command,
A steamboat few could beat,
Assembled on her barricade,
The sealers bright and gay,
With ringing cheers that rent the air,
As she sailed out that day.

She turns the heads at Eight O Clock,
And disappears from view
While fervent prayers for safe return,
Is offered for that crew,
On St. Patrick's Day they strike the seals,
The first boat in the fat,
When thirteen thousand prime young harps,
Fall from the sealers bat.

But on the twenty first of March,
A snow storm did arise,
That proved a hard and trying time,
On those poor sealer boys.
For on the Ice that cold March night,
Exposed to cold they stay,
While many poor fellows did succumb,
Before the dawn of day.

When forty eight poor christian souls,
Hard for their lives did fight,
From cold and hardship could not stand,
That fearful cold March night,
In light of glory they may see
It is the prayer of all,
Like Martyrs suffered for there sins,
Who died at duty's call.

Reprinted from THE HARBOUR GRACE STANDARD.
Harbour Grace, 1898.

The Greenland Disaster (6)

Off to the cold and stormy North,
Where all is seldom peace,
The blasts in wintertime descend,
The Storm King does not cease.
To there, off speeds the gallant ship
With brave men dead to fear
The day is fine, friends left behind,
Yet little do they care.

The icy fields at length they reach;
The day is clear, yet cold;
And coming night its shadow spreads
Over that band so bold.
Down, came the storm with giant force
Upon a frozen ocean;
They kneel to pray for coming day,
In heartfelt devotion.

Wearied all day from heavy toil,
Poor fare, and piercing cold;
With raging storm, no shelter made,
They suffered pains untold.
It was a dreadful night for all,
Each trusts to one another;
They try to cheer, yet in despair,
Some poor half-frozen brother.

And watching till the latest breath
Had left that frozen form;
Then turns a sad bewildered look
In deep and wild alarm;
Ah! mournful sight indeed to see,
It made the boldest shudder;
Those once so bold, benumbed with cold,
Fall beside another.

All night ring on the frosty air
Cries of helpless dying;
A drowning comrad not far off,
That same cry replying;
That pleading cry ascends to Heav'n
Just like incense of old;
Asks a blessing, rescues missing,
And stays the piercing cold.

Cease! ye fierce winds of heaven, cease;
And stay the piercing blast;
Haste! thou approaching morning, haste;
Your floods of daylight cast.
Bring comfort to the dying ones,
Assuage their dreadful pains,
And bring relief to stay their grief,
They cannot long remain.

The morning breaks with piercing shrieks;
Snow-clouds rent asunder;
A word goes round, "The ship in sight."
Caught dying ears like thunder.
Rousing from their snow clad beds
Full their reason failing,
Falls back again, there to remain,
Efforts unavailing.

The strongest then renewed with hope
Wearily plods along;
The weak, too tired themselves to stir,
Come guarded by the strong.
In ones and twos they reach the ship,
Night casts its mantle o'er,
Ah such a night: such awful sight
They had not seen before.

How does it fare with those now out?
Not one alive can know
The sorrows and the sufferings
They had to undergo.
They stripped the dead to clothe themselves,
They prayed, they sang, they cried;
Till one by one they all succumb,
And lying down they died.

The third day when the storm had ceased,
What few survivors came,
Were so benumbed with piercing cold
That most were near insane.
Assisted then from ships near by
They searched the ice around,
Till twenty-five, not one alive,
Dead bodies had been found.

Yet, scarcely knowing who were gone,
The roll began to call,
The rescued and the missing ones
Make forty-eight in all.
Cast down with care and lonely grief,
All hands feels now so sad;
The time so brief, that brought the grief,
Where all, but now, were glad.

A floating tomb of human woe
She leaves that fatal spot;
They think of those now left behind
Which cannot be forgot.
She reaches port a ship of grief,
Gloomy, sad, and shockin';
The Captain he, 'tis sad to see,
Could not be forgotten.

Why need we mourn for those now left
Beneath the angry wave;
They sleep as well beneath the sea
As others in their grave.
And on that dreadful judgment day
The sea will give its dead;
Those in the deep shall rise that sleep,
The word of God hath said.

As for their souls how can we tell;
For God would hear our cry,
If from our hearts forgiveness ask
Our hour before we die.
And those who died with hearts bowed down,
Were surely touched with grief,
Their sins forgiv'n, they reach Heav'n
Much like the dying thief.

A. House
Ship Cove, T.B., April 20, 1898

Reprinted from THE HARBOUR GRACE STANDARD.
Harbour Grace, 1898.

The Greenland Disaster (7)

We now relate the mournful fate of gallant sealers dead,
Whose frost-burnt hands and faces paid the price of daily bread;
A vanished race of experts on the rafting Arctic floe:
The victims of disaster in a wilderness of snow...

On March the tenth in Ninety-eight, their sealing-ship departed
With blaring horns and lusty cheers from sealers eager-hearted;
Then no one knew that such a crew would pay grim sacrifice
To demon spirits ruling o'er the realms of Polar ice.

A dauntless master-mariner from Bonavista Bay,
George Barbour, brought the *Greenland* where the whining whitecoats lay;
And ninety miles north-east the Funks they struck the young harp patches.
Then cruised about to kill and pan and load her to the hatches.

Upon the fateful twenty-first, the weather-glass was low;
The dawning brought a leaden sky, presaging wind and snow;
But bred to battle wintry storms, these vet'rans of the North,
At duty's call, took Danger's risk and bravely ventured forth.

The Polar legions hurled their wrath against the hapless men
Who struggled through the blinding snow to reach their ship again: —
But forty hours passed before they knew the tragic score;
That forty-eight had met their fate, to hunt the seals no more!

They'd never seen a wilder storm than that which raged that day;
So fierce and swift the stifling drift, they wandered far astray;
Exposed to piercing icy blasts upon that frozen waste,
The only shelter they could find was in a friend's embrace.

Aboard the *Greenland*, frantic mates were also horrified
When pressure of the hurricane careened her on her side;
All night they worked like Trojans, shifting cargo on the deck.
To put her on an even keel, preventing certain wreck.

The foil'wing day, at 4 P.M., improvement in the storm
Permitted rescue-squads to bring them food and clothing warm;
In contemplation, strong men wept, as others would relate
The tragic circumstances of their comrade's gruesome fate.

A sister-ship, *Diana*, and the *Iceland* joined the search
To find the dead and living and retrieve them from their perch
Upon the fatal pans of ice. Then on the twenty-third
All victims and survivors to the *Greenland* were transferred.

Imagine frozen corpses, packed in ice, upon a ship;
A floating-Horror-chamber was the *Greenland* on that trip;
For twenty-five stiff carcasses are not a pleasant sight
To muse upon in solitude when storms are at their height.

The fiendish storm-king, unappeased, assaulted, yet once more,
The *Greenland* ere she reached St. John's, and drove her on the shore,
Off Bay de Verde, the crew engaged in shifting bunker coal
Till, leaving half her keel behind, she floated off the shoal.

In multitude the city-folk looked on with bated breath
As slowly in the harbour steamed that tragic ship of death.
The Premier and the Governor; the Legislature too
In silence waited on the pier to greet the mournful crew.

The melancholy story of the *Greenland's* grim disaster
Was soon officially received in sorrow from her master:
Survivors gave their narratives in detail to the Press,
While tearful eye and broken voice displayed the soul's distress.

A scene to shock the stoutest heart then met the eyes which gazed
Upon the dead as from the hold the frozen forms were raised;
All warped beyond resemblance to their former nimble frames,
Those twisted shapes were all that then remained of sealers' names.

But human names are nothing, if with nerve and spirit they
Are not endowed as were these men we celebrate today.
When high adventure challenges, the dauntless heart replies
With noble deeds of daring and, heroic, lives — or dies.

In tribute now, our heads we bow for valiant sealers dead;
Whose frost-burnt hands and faces dearly paid for daily bread;
The forty-eight who met their fate amid the ice and snow
While battling raging blizzards on the rafting Arctic floe.

(Won second prize in Ballad Section of Government sponsored Arts and Letters Contest, March 1961.)
(Original note)

The Ice-Floes

Dawn from the Foretop! Dawn from the Barrel!
   A scurry of feet with a roar overhead;
The master-watch wildly pointing to Northward,
   Where the herd in front of *The Eagle* was spread!

Steel-planked and sheathed like a battleship's nose,
She battered her path through the drifting floes;
Past slob and growler we drove, and rammed her
Into the heart of the patch and jammed her.
There were hundred of thousands of seals, I'd swear,
In the stretch of that field — "white harps" to spare
For a dozen such fleets as had left that spring
To share in the general harvesting.
The first of the line, we had struck the main herd;
The day was ours, and our pulses stirred
In that brisk, live hour before the sun,
At the thought of the load and the sweepstake won.

We stood on the deck as the morning outrolled
On the fields its tissue of orange and gold.
And lit up the ice to the north in the sharp,
Clear air; each mother seal and its "harp"
Lay side by side; and as far as the range
Of the patch ran out we saw that strange,
And unimaginable thing
That sealers talk of every spring —
The "bobbing holes" within the floes
That neither wind nor frost could close;
Through every hole a seal could dive,
And search, to keep her brood alive,
A hundred miles it might well be,
For food beneath that frozen sea.
Round sunken reef and cape she would rove,
And though the wind and current drove
The ice-fields many leagues that day,
We knew she would turn and find her way
Back to the hole, without the help
Of compass or log, to suckle her whelp —
Back to that hole in the distant floes,
And smash her way up with her teeth and nose.

But we flung those thoughts aside when the shout
Of command from the master-watch rang out.

Assigned to our places in watches of four —
Over the rails in a wild carouse,
Two from the port and starboard bows,
Two from the broadsides — off we tore,
In the breathless rush for the day's attack,
With the speed of hounds on a caribou's track.
With the rise of the sun we started to kill,
A seal for each blow from the iron bill
Of our gaffs. From the nose to the tail we ripped them,
And laid their quivering carcasses flat
On the ice; then with our knives we stripped them
For the sake of the pelt and its lining of fat.
With three fathoms of rope we laced them fast,
With their skins to the ice to be easy to drag,
With our shoulders galled we drew them, and cast
Them in thousands around the watch's flag.
Then, with our bodies begrimed with the reek
Of grease and sweat from the toil of the day,
We made for *The Eagle*, two miles away,
At the signal that flew from her mizzen peak.
And through the night, as inch by inch
She reached the pans with the "harps" piled high,
We hoisted them up as the hours filed by
To the sleepy growl of the donkey-winch.

Over the bulwarks again we were gone,
With the first faint streaks of a misty dawn;
Fast as our arms could swing we slew them,
Ripped them, "sculped" them, roped and drew them
To the pans where the seals in pyramids rose
Around the flags on the central floes,
Till we reckoned we had nine thousand dead
By the time the afternoon had fled;
And that an added thousand or more
Would beat the count of the day before.
So back again to the patch we went
To haul, before the day was spent,
Another load of four "harps" a man,
To make the last the record pan.
And not one of us saw, as we gaffed, and skinned,
And took them in tow, that the north-east wind
Had veered off-shore; that the air was colder;

That the signs of recall were there to the south,
The flag of *The Eagle*, and the long, thin smoulder
That drifted away from her funnel's mouth.
Not one of us thought of the speed of the storm
That hounded our tracks in the day's last chase
(For the slaughter was swift, and the blood was warm),
Till we felt the first sting of the snow in our face.

We looked south-east, where, an hour ago,
Like a smudge on the sky-line, someone had seen
The Eagle, and thought he had heard her blow
A note like a warning from her siren.
We gathered in knots, each man within call
Of his mate, and slipping our ropes, we sped,
Plunging our way through a thickening wall
Of snow that the gale was driving ahead.
We ran with the wind on our shoulder; we knew
That the night had left us this only clue
Of the track before us, though with each wail
That grew to the pang of a shriek from the gale,
Some of us swore that *The Eagle* screamed
Right off to the east; to others it seemed
On the southern quarter and near, while the rest
Cried out with every report that rose
From the strain and the rend of the wind on the floes
That *The Eagle* was firing her guns to the west.
And some of them turned to the west, though to go
Was madness — we knew it and roared, but the notes
Of our warning were lost as a fierce gust of snow
Eddied, and strangled the words in our throats.
Then we felt in our hearts that the night had swallowed
All signals, the whistle, the flare, and the smoke
To the south; and like sheep we huddled and broke.

To the south; and like sheep in a storm we followed
Each other; like sheep we huddled and broke.
Here one would fall as hunger took hold
Of his step; here one would sleep as the cold
Crept into his blood, and another would kneel
Athwart the body of some dead seal,
And with knife and nails would tear it apart,
To flesh his teeth in its frozen heart.
And another dreamed that the storm was past,
And raved of his bunk and brandy and food,
And *The Eagle* near, though in that blast
The mother was fully as blind as her brood.

Then we saw, what we feared from the first — dark places
Here and there to the left of us, wide, yawning spaces
Of water; the fissures and cracks had increased
   Till the outer pans were afloat, and we knew,
As they drifted along in the night to the east,
   By the cries we heard, that some of our crew
Were borne to the sea on those pans and were lost.
   And we turned with the wind in our faces again,
   And took the snow with its lancing pain,
Till our eye-balls cracked with the salt and the frost;
Till only iron and fire that night
   Survived on the ice as we stumbled on;
As we fell and rose and plunged — till the light
   In the south and east disclosed the dawn,
And the sea heaving with floes — and then,
The Eagle in wild pursuit of her men.

And the rest is as a story told,
   Or a dream that belonged to a dim, mad past,
Of a March night and a north wind's cold,
   Of a voyage home with a flag half-mast;
Of twenty thousand seals that were killed
   To help to lower the price of bread;
Of the muffled beat...of a drum...that filled
   A nave...at our count of sixty dead.

Reprinted with permission from E.J. Pratt, HERE THE TIDES FLOW, (Toronto, MacMillian Company of Canada Limited, 1962), 58.

Sealing Fifty Years Ago

"Four hundred sail of shipping fine
Could then be seen at anchor,
Awaiting time to fall in line
And for a sou'west spanker.
With topsail's set, yea, every sail,
And brave men, stout and jolly,
To face a strong, bold northern gale
With Terry or John Holley.

Six hundred thousand seals were caught
'Midst northern fields of slaughter,
The greatest number ever got
And brought safe o'er the water.
But that was fifty years or more
And half that catch today
Is now brought safely back to shore
With hearts not half so gay."

Reprinted with permission from Michael Murphy, PATHWAYS THROUGH YESTERDAY. Edited by Gerald Moore. (St. John's: Town Crier Publishing Company Ltd., 1976), 158. This song by James Murphy was originally published in Johnny Burke and James Murphy, THE DUKE OF YORK SONGSTER AND CHRISTMAS ADVERTISER. (St. John's: Burke & Murphy, 1901).

The Sealers' Strike of 1902

Attention, all ye fishermen, and read this ballad down,
And hear about the sealers strike the other day in town;
When full three thousand northern men did walk the streets all day,
With cool determined faces they struck out to get fair play.

Each steamer's crew did fall in line, while cheers out loudly rang,
Led on by one brave Calloway, the hero of the gang.
Free berths it was their motto, and no man would give in,
A fight for death or glory, boys, this victory to win.

They halted just before the bank, when all hands fell in line;
They went inside to state their case before A.B. Morine.
He got the terms to suit the men, and from the van did call;
That he secured three fifty and "free berths" for one and all.

A ringing cheer the sealers gave, with hearts both light and gay.
And three more cheers they gave Morine, the man who won the day.
With happy hearts they fisted bags, as lightly they did trip,
With boots and bags and baking pan to get on board their ship.

Then soon around the northern head they disappeared from view,
Manned by a plucky, hardy race, a bully northern crew.
May they return with bumper trips, it is our earnest prayer
The boys who nobly showed their pluck, and fought to get their share.

Three thousand were on strike. Free berths and a rise.

The ever to be remembered sealers strike took place in St. John's on Friday, March 8th, 1902. Three thousand sealers refused to proceed to the ice-floes until their demands were complied with. The strike ended March 12th after the sealers demands were granted. The leader of the strikers was one Calloway, who died a year or so afterwards. This song was written by John Burke, the bard of Prescott Street.
(Original Note)

Reprinted with permission from James Murphy, SONGS SUNG BY OLD TIME SEALERS OF MANY YEARS AGO. (St. John's: James Murphy, 1925), 20.

The Sealers Gained the Strike

Attention! all ye fishermen,
  And read this ballad down,
And hear about the sealers' strike
  The other day in town;
When full five thousand northern men
  Did walk the streets all day,
With cool determined faces
  Struck out to get fair play.

Each steamer's crew did fall in line,
  While cheers out wildly rang,
Led on by one brave Colloway
  The leader of the gang.
Free berths, it was their motto boys,
  And no man would give in,
A fight for death or glory boys,
  This victory to win.

They halted just before the bank
  When all hands fell in line,
They went inside to state their case
  Before A.B. Morine.
He got the terms to suit the men,
  And from the van did call,
That he secured Three-fifty,
  And free berths for one and all.

A ringing cheer the sealers gave,
  With hearts both light and gay,
And three times three they gave Morine
  The man who gained the day.
With happy hearts they fisted bags,
  As lightly they did trip,
With boots and bags and baking pans
  To get on board their ships.

When soon around the northern head
  They disappear from view,
Manned by a plucky, hardy race,
  A bully northern crew.
And may they come with bumper trips
  It is our earnest prayer,
The boys who nobly showed the pluck
  And fought to get their share.

Reprinted with permission from John White, BURKE'S BALLADS. (St. John's: privately published, 1960), 14.

The Luck Went With the Sealers Since Brave Colloway Led the Strike

Attention, boys of Newfoundland,
  Who plough the raging main,
And hear about this sealers' strike,
  That I will now explain.
Ye struck, my boys, for better terms,
  The terms that ye did like;
And this is why brave Colloway,
  Did lead the sealers strike.

Three-fifty was the figure boys
  And nothing less would take,
Ye walked the streets with little food,
  No drink your thirst to slake.
And no man would give in an inch,
  Though hunger's pangs they knew,
Until their terms did satisfy
  Both every steamer's crew.

Up spoke the leader of the strike,
  Come boys, now fall in line,
We'll go at once and state our case
  To Alfred B. Morine;
And he will see we get fair play,
  Spoke Colloway loud and clear,
And from five thousand manly throats
  Went up a deafening cheer.

He called at once on Mr. Baird,
  On Mr. Job and Grieve;
Who said, of course we'll give the terms,
  And no man will deceive.
The scene that day on McBride's Hill,
  When Morine took the van,
Will ever hang on memory's wall
  In every boy and man.

The luck went with the sealers boys,
  Though late they sailed away;
They struck the white coats in their prime
  Upon St. Patrick's Day.
The best of seals the ships brought in.
  And clever bills laid down,
A year to be remembered
  When the strike was in the town.

Ice Bound Hunting Seals

The wind was still from the nor' east
As we sat down to our humble feast,
In the forecastle of our good old ship
And muttered as we took a sip;
"The days gone by."

Contrasted this with years long past,
With Captain Mick* our luck we cast;
And how we worked with heart and will
And seldom came without a "bill"
In the Argo.

Then Jack spoke up and told his yarn,
How Captain John at break of marn
Would down the forecastle loudly bawl:
"Come, harry boys, I hear them call,
To the eastward."

And how to Islands he would moor,
How many times he filled the Suir
"In such a place," spoke Captain Dan,
"And he was just the very man
To fill her up."

Then spoke Bill as he heaved a sigh,
"Let's think of these, not times gone bye;
They're now all gone and can't be helped,"
He looked as what he said he felt —
And dropped it.

Then uncle "Garge," the oldest man,
Said "he for one didn't care a 'darn',"
He knowed it well, and with a scream
Cried out "it's the 'infarnal' steam —
That's what done it."

*Fitzgerald

Reprinted with permission from an unidentified manuscript, believed to be the work of James Murphy, and deposited in the Memorial University of Newfoundland Folklore and Language Archive (hereafter cited as MUNFLA) by Dr. John Widdowson of the University of Sheffield. Courtesy of Fred Earle, Memorial University Extension Service.

Four sealers at work c. 1900. Two men are "sculping" another is uncoiling his tow rope and a fourth man is holding the ship's flag which will be used to mark the pile of seal pelts. Note the upright gaff — the killing tool — which is no longer permitted to be used.

Courtesy of the Provincial Archives of Newfoundland and Labrador (PANL).

The Sealer Lad

The sealer lad from his home is gone
On board his ship you'll find him.
His sheathed knife he has buckled on
And his clothes bag slung behind him.
Frozen jam, said the hunter bold,
My tongue shall ever praise thee
I long to tread thee once again
Tho' the venture hardly pays me.

In days gone by when flint lock guns
Were reckoned things of beauty,
Ship load of harps meant a handsome bill
Now it scarce pays Alfred's duty.
Still I long to tread the frozen pans
I yearn, the floe, to greet thee
On thy kind breast my whitecoats rest
When the steamships shall meet thee.

Ere the moon shall wane the sealer's heart
Will throb, I know with pleasure,
And the depth of fat on the young harps back
His hunting knife shall measure.
No flag he'll plant on the frozen pans
He'll leave but blood-red patches
No more he'll pan to enrich the man
Of a thousand inky scratches.

Reprinted from John Burke and George Oliver, THE PEOPLE'S SONGSTER, BUYERS' GUIDE AND GEMS OF POETRY AND PROSE. (St. John's: Burke and Oliver, 1900), 30.

When The Flippers Strike The Town

You may talk about the pan-cakes
That your mother used to fry,
The ones that you remember
When you were a little boy
But this I got to tell you
And upon me do not frown
The pan-cakes won't be "in it"
When the flippers "strike the town."

In the morning you'll rise early
And you'll need no Daylight Bill
When a "swiler" first is sighted
From the tower upon the "Hill"
And the latest married woman
As she sports her wedding-gown
She'll be smiling on the "swilers"
When the flippers "strike the town."

Now there's nothing like "a flipper"
No there's not — "upon my eye"
Fried or stewed or "parboiled"
Or well baked up in a pie.
Sure you wouldn't view a comet
If its tail lay on the 'grown'
Upon my word and honor
When the flippers "strike the town."

When the father of the family
To his wife will bring a "gad"
How her eyes will gaze upon them
Oh, how pleased she'll be and glad
And she'll talk about the frosty spring
When he made forty poun'
Pork and cabbage won't be in it
When the flippers "strike the town."

Arrival of the "Grand Lake" and "Virginia Lake" With Bumper Trips

The *Grand Lake*, boys is coming in
With bunting waving grand,
Manned by a crew of hardy lads,
Who belong to Newfoundland.
With twenty thousand prime young harps,
She had them stowed below,
And more in pounds upon her deck
Just from the northern floe.

She struck the seals with all the rest,
Her crew worked with a will,
When thirty thousand prime young harps
Those northern boys did kill;
She then steamed round, picked up her pans,
To hoist on board began,
While willing hands soon filled her deck,
Good work for every man.

All hands on board she left for home,
A dandy trip of fat,
That fell boys to the *Grand Lake's* crew,
Killed by the sealers bat.
And three loud cheers for Henry Dawe,
Likewise his gallant crew,
And may he live for many springs,
The whitecoats for to slew.

The old *Virginia*, filled with fat,
Soon followed in her wake,
With five and twenty thousand harps
She from the patch did take;
For William Winsor, did it well,
The leader of the fleet,
Besides a bully crew of men,
No steamer's crew can beat.

Well done the sealers, bully crews,
Ye did it well this spring,
While gaily on the barricade,
With happy hearts you sing.
The *Grand* and old *Virginia Lake*
Ye filled them up in style,
The hardy sons of Newfoundland
Who at the ice must toil.

Reprinted from the OLD HOME WEEK SONGSTER, (1904), 21. Courtesy of Paul Mercer.

First Arrival — "Aurora" and "Walrus" Full

The first arrival from the ice
  Has just come in to-day,
The good old ship *Aurora*,
  And her colors waving gay;
With four and thirty thousand seals
  She had them stowed below
Besides five thousand on her deck.
  Right from the northern floe.

On St. Patrick's Day in spirits gay,
  They struck the heavy patch
And seventeen thousand prime young seals
  They had below her hatch;
And all next day lay in the patch,
  Where they worked with a will
Until the good sealer, boys,
  With prime young harps did fill.

Well done the boys from northern bay
  Ye always make a name,
The old *Aurora*, always fill,
  With Captain Abram Kean,
And may ye live for many years,
  This bonnie sealing crew,
To be the jowlers of the fleet,
  The whitecoats for to slew.

The staunch old *Walrus* now is in
  The second of the fleet,
With sixteen thousand prime young harps
  And better ships did beat
It was plucky Capt. Winsor boys
  The *Walrus* well did fill,
Manned by a crew of hardy lads,
  Belonging to Wesleyville.

Reprinted from the OLD HOME WEEK SONGSTER, (1904), 23. Courtesy of Paul Mercer.

Arrival of "Aurora," "Diana," "Virginia Lake" and "Vanguard," Loaded

All welcome to the northern fleet,
  That just arrived to-day,
Pounds filled up with prime harp seals
  And colours waving gay.
While sealers on the barricade
  Indulge in friendly chat,
While round the northern head they steam,
  Filled up with prime young fat.

The first to steam around the heads
  Is Captain Abram Kean,
Who holds the record at the ice
  And well deserves the name.
Her crew of hardy northern boys
  Did haul on board their tow,
When soon they filled the good ship up
  As much as she could stow.

Then Captain Barbour signalled next,
  Bound on for St. John's town,
The lucky old *Diana*
  In the water is well down.
With all the rest quite near the Funks
  They stuck the harp and pup,
Her bully crew worked night and day
  Until they filled her up.

Close in their wake, at early morn,
  The *Virginia Lake* we see,
Manned by the jowler of the fleet
  The famous Captain Knee.
Her crew of Terra Nova's sons
  A cleaver bill will win,
With two and twenty thousand seals,
  No better yet came in.

Well done the *Vanguard* and her crew,
  You did the dog-hood face,
When Captain Barbour brought her, boys,
  In full to Harbour Grace.
May health, wealth and prosperity
  Fall on you where ye go,
Brave sealer boys to spare
  To face the northern floe.

Reprinted from an unidentified manuscript presented to MUNFLA by Mr. Leo Moakler (n.p.,n.d.). In all probability the work is that of Johnny Burke. Courtesy of Paul Mercer.

The Nimrod's Song

Come all ye friends of Newfoundland
Who have a mind to roam,
O'er the wild and stormy ocean
Far from your native home;
I hope you'll pay attention
And listen to my song.
To those few lines I'm going to sing
It won't detain you long.

It was on the 20th day of March
In the ice we got jammed.
Between the island of St. Paul's
And the shores of Newfoundland.
The wind being from the South South East
All mixed with showers of snow,
And each man packed his clothes, boys,
And from her he did go.

The captain, mate and boatswain
Stood on the bridge that time,
Some of our crew went down below
Provisions for to find.
It would grieve the hearts of any man
To see us standing so
Upon the frozen pans of ice
Amidst the frost and snow.

The ice still kept on rafting
Our ship still going down
Until by the Providence of God
The ice it wheeled around.
Our captain he stood on the bridge.
The ice it gave a slack.
And by the moving of his hand
He ordered all hands back.

Some of us tried to board her,
But it was all in vain,
The force of ice came bundling in
Her decks we could not gain.
Some of our crew looked sorrowful
On board of us that day
Until the ice it gave a slack
And we got underway.

Jim Barrett's been our deck master,
The berth that he did fill,
He climbed the rigging of the ship
To cut away the sail.
He cut the mainsail from the mast,
As you may understand,
For to make a covering for us
Out upon the frozen jam.

The ice it broke our quarter boats,
Like wise our quarter sail,
It didn't hurt our steering gear
Nor hinder us from sail.
Our captain been a hero bold,
A man of heart and skill,
his name is Baxter Barbour
Belonging to Wesleyville.

John Gibbon's been our carpenter,
The same I will relate.
"Fred" Newbury our boatswain,
"Bill" Thistle boatswain mate;
Our cabin cook from Carbonear,
James Poole it is his name,
Saved nothing but his cooking suit
To face the icy plain.

George Snow he was our cabin steward.
The same I will pen down,
Our mate been Thomas Rideout,
A native of New Town;
John Crocker was our for'ad cook,
"Paddy" Murphy does the same,
And the man that minded the lobby,
Mark Norris is his name.

Success to Captain Barbour,
And may he soon command
A better ship than *The Nimrod*
And a crew from Newfoundland.
May good luck too attend him
While on the northern seas,
And may his big jib always draw
Filled by a moderate breeze.

Reprinted with permission from James Murphy, SONGS SUNG BY OLD TIME SEALERS OF MANY YEARS AGO. (St. John's: James Murphy, 1925), 1.

Captains and Ships

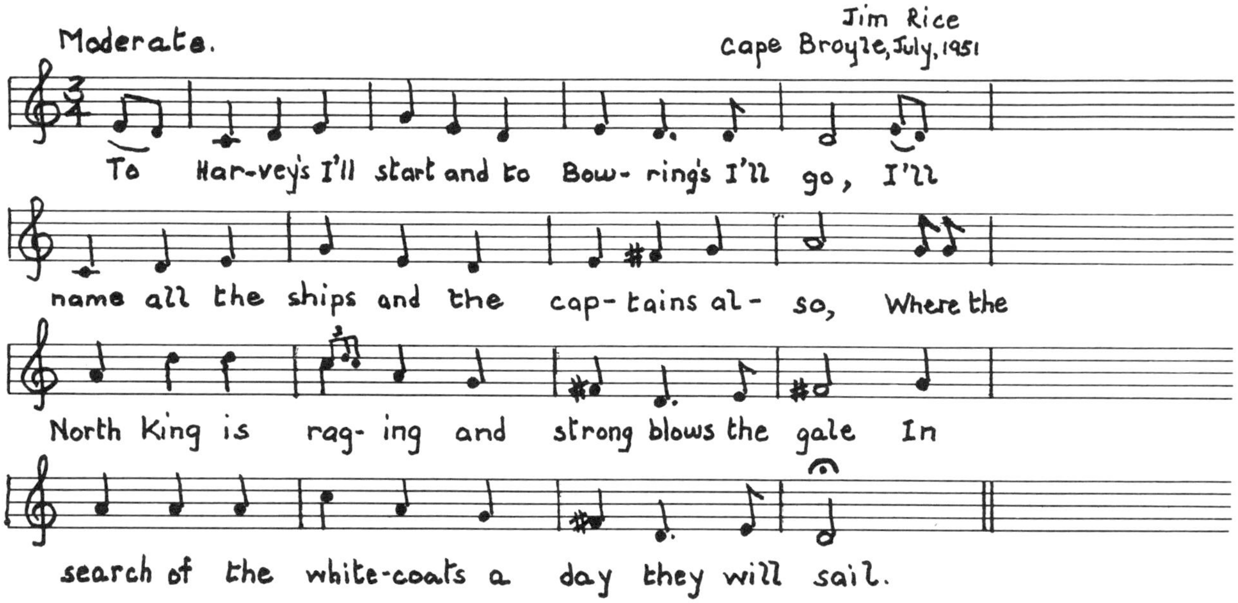

To Harvey's I'll start and to Bowring's I'll go,
I'll name all the ships and the captains also,
Where the North King is raging and strong blows the gale
In search of the white-coats a day they will sail.

In the *Ad*, Captain Doyle; in the *Belle*, Joby Knee;
In the *Bon*, Captain Parsons, a stout man is he.
And jolly gay Kean in the spring will command
Harvey's port steamer the old *Newfoundland*.

The name of the ships of Job Brothers and Co.
The first and the next one I'll let ye all know;
The good ship *Diana* Joe Blandford has charge,
And I hope she'll come back before long with a surge.

In comes the *Beothic* so swift and so sure,
I hope that George Barbour is with her once more,
And then there's the *Neptune*, I see it so plain,
And likewise the *Eric* with pleasant Joe Kean.

I wish them good luck and from Job's take my leave,
I'll name all the ships owned by Walter Bain Grieve;
There's the *Bloodhound*, Gate Windsor, he'll yet make his mark,
With Kean in the *Iceland*: the *Cross* Captain Clark.

Their brothers also got a fine man for sure,
Baxter Barbour that sails in the ship *Labrador*,
He'll fill 'er and bring 'er safe over the seas
As he did when he sailed in the Crosbie's *Louise*.

Now and then it is Bowring's the firm that's well-known
For the pluck, push and enterprise and all it has shown
By building that steamer, ye all know her well,
Evan Kean her commander, the new *Florizel*.

The *Kite*, Captain Carrol, I wish him good luck,
Once more in the *Eagle* Joe Kean showed his pluck;
And Bartlett, the *Viking* I pray ye will fill,
And dear old Dan Green, may he make the big bill.

The *Ranger*, Sam Windsor be fleetingly seen,
Likewise Noah Bishop and the *Algerine*.
My song is concluded 'bout captains and ships,
May they all come in with big beards on their lips!

Men returning to the ship with their "tows" of seals, c. 1900.

Courtesy of the Provincial Archives of Newfoundland and Labrador (PANL).

The Sealer's Song

The Terra Nova, Captain Kean
with two hundred and three men,
Went through the gap this morning
To try their luck again.

The screaming of the *Eagle*
was heard both high and low,
Commanded by a son of Abraham
That sturdy Capt. Joe.

Next came the stout ship *Aurora*
Commanded by Dan Green
And Capt. Bob Winsor
In Bowring's *Algerine*.

She aspied a bold-eyed stranger
All on her sealing quest
No notice to her danger
With a Bishop she is blessed.

Darius Blandford sails the *Vanguard*
The pride of Baine Johnston's ships
Long live Capt. Blandford
To bring in bumper trips.

There is another that can find them
Where e're there to be found
That's bulley young Bill Windsor,
Who sails the old *Bloodhound*.

Joe Barbour sails the *Iceland*
A bold seal killer too,
Bragg commands the *Greenland*,
To show what Bragg can do.

Here's luck to Peter Carter
And may he meet no loss
You won't find one no smarter
For to sail the *Southern Cross*.

George Barbour sails the *Neptune*
He's as good as water wet
Jobie Kean commands the *Eric*
He's a solid man you bet.

That lucky ship *Dianna*
She bears a man of fame
From Newtown, Bonavista,
Alpheus Barbour is his name.

The *Adventure* and the *Grand Lake*
Is also gone to sea
The first on charge is Harry Dawe
The latter by Jobie Knee.

John Parsons in the *Newfoundland*
Sent forth his whistles shrill
Jessie Winsor sails the *Panther*,
He's the pride of Wesleyville.

Jacob Winsor sails the *Walrus*
Supplied by Bowring's store,
Our good friend Baird and Gordon
Sent George Hand in the *Labrador*.

Jacob Kean commands the *Virginia Lake*
Makes up the twenty ships
And my best wish may they all return
Filled up with bumper trips.

Johnny Burke

Reprinted with permission from the private collection of the late Gerald S. Doyle.

Six Men and One Woman Taken Off The Ice At Petty Harb'r

Come all you hardy Fishermen,
And hark to what I say,
And hear how six were rescued boys
Near Petty Harbor Bay.
And one lone woman with the rest
Upon the Ice all night.
Till early Sunday morning,
When the *Ingraham* hove in sight.

A Schooner owned by Capt. Dicks,
Set sail about sunrise,
Bound home before a pleasant breeze,
With Fishery supplies.
When near Shoal Bay, they met the ice,
That forced her near the land.
The crew was forced to take the ice,
The strain she could not stand.

All night upon the open ice,
These poor men had to stay,
Until they spied the Tug Boat,
Just about the break of day,
While some got on a shelving rock,
From boistorous seas ran high,
Not to remain upon the ice,
On that cold night to die.

But soon the news spread round boy
And very soon reached town.
When soon the *Ingraham* left St. John's,
And on the crew bore down,
When six men and one woman,
They landed safe on shore,
Picked up near Petty Harbor Bay,
Kind friends to see once more.

Reprinted from THE HARBOUR GRACE STANDARD.
Harbour Grace, March 13, 1909.

Success to the Hardy Sealers

The 12th of March is drawing near
  And we must all prepare
Our pipers and our pannicans,
  The sealer's life to share —
And in the ball-room build our bunks,
  Until we burn her down —
And hope to come with bumper trips,
  All safe to St. John's town.

Some in the gulf will try their luck,
  The white-coats for to fight —
The *Algerine* and *Southern Cross*,
  The *Neptune* and the *Kite*,
The *Viking* and the *Labrador*
  And all the wooden ships —
May they return all safe and sound
  With surging heavy trips.

The *Venture* and the *Beothic*
  And the ships we know so well
Are manned by sons of Newfoundland
  Likewise the *Florizel*,
The *Stephano* and the ironclads
  Will face the Northern floe,
With strapping lads from Northern bays
  Who well can haul a tow.

The time is drawing handy,
  When the ships will sail away;
And thousands lined on every pier
  To watch the ships that day —
And wishing all a pleasant voyage
  And every steamboat fill,
To pay the boys that hauled the seals
  And make a tidy bill.

Then here's success to all the fleet,
  That on the twelfth will sail,
May they return with bumper trips
  And filled up to the rail.
And soon arrive in port again,
  Safe from the Northern seas,
With bunting decked on every spar
  And flying in the breeze.

May ye return all safe and well,
  All in good health and cheer,
Safe to your homes in Newfoundland,
  To friends you love most dear.
We hope to hear good news from all,
  And fervently we pray,
That every steamer has her load,
  Before St. Patrick's day.

Reprinted from Johnny Burke, BURKE'S BALLADS. (St. John's: Johnny Burke, 1912), 14.

First Arrival from Seal Fishery S.S. Fogota, 1912

The first arrival from the front,
Is just come in to-day,
The little ship *Fogota*,
With her colors waving gay
And with a load of prime young harps,
As to her pier does sail,
Her hardy crew of Northern boys
Are leaning o'er the rail.

The *Fogota* was the first to sail,
Late in the afternoon,
And shaped her course to strike the patch,
Where she did meet them soon.
When near a place they call the Funks
As she was steaming slow,
Her crew, they killed nine hundred,
And had them stowed below.

The main patch they did boldly strike,
Her crew in spirits gay,
And thirty hundred prime young harps,
They killed on Sheelah's day.
The *Ranger* was the only ship,
The *Fogota's* men could see,
And soon she steamed among the seals,
In charge of Captain Knee.

Next day they killed two thousand more
And stowed them all below,
The weather, it was stormy
And a north-east wind did blow,
She picked up scattered seals next day,
And then for home she bore,
For the Captain he decided
That the ship could take no more.

With colours flying to the breeze,
And filled up to the rail,
The little ship *Fogota*
Round the Northern Head did sail.
While thousands watched her from the shore
On every wharf and pier,
To greet the first arrival
And to give a lusty cheer.

Success to Captain Winsor
And his crew of Northern boys,
They'd fill another steamer up,
If she was twice her size.
And may you make a tidy bill,
It is the wish of all,
The *Fogota's* hardy sealers,
The white coats they did haul.

Reprinted from Johnny Burke, BURKE'S BALLADS. (St. John's: Johnny Burke, 1912), 20.

Captain Bob Bartlett

A rugged Newfoundlander as ever sailed the seas.
He was born and raised at Brigus in the bay,
Where Conception's water ruffles to the humming of each breeze,
And August brings the scent of new mown hay.
From the time he left his cradle he hankered for the sea,
As around him famous captains formed a hub.
He distracted maids and mother at the tender age of three,
When he sailed the flooded kitchen in a tub.

Then forward to a sealer's deck on seaman's feet he stood,
Where he thrilled to hear the master watch's call:
Come, every mother's son of ye! take ropes and knife, and food,
Then conclude yer prayers out where the whitecoats bawl.
Now Bob became a captain where he earned a dashing name,
On schooners, and in every skiff and punt.
Onboard all sealing steamers the name enlarged to fame,
Both in the Gulf and down the Northern front.

And then when Admiral Peary had crave to seek the Pole,
He selected Captain Bartlett for his guide.
And Peary frankly stated he would not have reached his goal,
Without that stalwart skipper by his side.
He walked as well on snowshoes where the old Aurora flashed,
As he did in patent leather on a floor.
He broke the trail for Peary when at the Pole they dashed,
And earned undying praise for evermore.

Then when that expedition rejoicing homeward came,
Bob Bartlett's voice was sought on every hand;
And he pictured his adventures in words he would not tame,
The salty blunt spoke words of Newfoundland.
His schooner *Effie Morrissey* became a household name,
As from Arctic expeditions she came forth;
And down the years which followed she enlarged her skipper's fame,
For she was his magic carpet to the North.

84

Though honour after honour was conferred on Captain Bob
He always kept in touch with Newfoundland;
For each summer when he left the States on some exploring job,
He never failed to visit on our strand.
In nineteen hundred and forty-six Bob Bartlett sailed away
On the greatest expedition of them all.
He's resting now at Brigus where his grave o'erlooks the bay,
Guarded by his mother's roses 'neath the wall.

Reprinted with permission from Otto Kelland, ANCHOR WATCH: NEWFOUNDLAND STORIES IN VERSE. (St. John's: Dick's & Co., 1960), 43.

Ballad of Captain Bob Bartlett, Arctic Explorer

Bob Bartlett, born in Brigus, of a bold sea-faring breed,
Became a master-mariner as destiny decreed:
He won renown by practicing the brave explorer's role
When Peary used his services to the Northern Pole.

The briny spray of Turnavik he faced in early youth;
While fishing with his kin-folk there, he learnt the ocean's truth;
Where wrestling with the ropes and lines, the windlass, oars and sails,
His strong physique developed to withstand the Arctic gales.

In early manhood days he showed great promise of that fame
Which history in later years accorded Bartlett's name.
He furnished for museums countless specimens worthwhile,
And scientific data for the Hydrographic file.

His little ship the *Morrissey* possessed his heart and mind,
As gracefully she sailed along, close-hauled against the wind;
He cheered her on with ev'ry wave which broke across her bow,
As, through the hissing foam, he watched his vessel swiftly plough.

The gulls which hovered over him in whistling Autumn gales,
The bull-birds, turrs, and kittiwakes — the porpoises and whales;
The bosun, bawk and curlew — the phalarope and plover
Were frolicsome companions of this North Atlantic rover.

The off-shore squalls, the rain and fog, the drizzle, sleet and snow —
The frothy-lipped North-easter, the close-packed Arctic floe:
The rocks and shoals, the treacherous tides, the rough forbidding shore
Present their dangerous challenges to Bartlett nevermore.

The sealing ships whose bloody decks betokened paying trips,
As ev'ry Spring they wriggled through the ice-field's crushing grips:
The *Nimrod*, *Neptune*, *Algerine*, — the *Bonaventure*, *Kite*
No more, with Bartlett on the bridge, shall pass the harbour light.

Makkovik, Ailik, Kiglaipait, Cape Harrison and Nain
Shall never see our Newfoundland adventurer again;
For now beneath his native sod there lies, in sacred trust,
A legacy immortal — Bob Bartlett's honoured dust.

Reprinted with permission from A.C. Wornell, RHYMES OF A NEWFOUNDLANDER. (St. John's: Guardian Limited, 1951), 11.

Maurice Crotty

Sit down while I sing you a ditty,
Of the spring I was out in the *Dan*;
Maurice Crotty was one of her sealers,
A comical cure of a man.

He could spin out a yarn by the hour,
And lies he could tell by the score,
And when Maurice came down in the ballroom,
All hands in 'er body would roar.

It was his first spring at ice hunting,
Not a rope in the ship did he know,
Not even to fold up a bunting,
And awkward to lace up a tow.

Then the captain sang out one fine morning,
"Come Crotty, your trick at the wheel."
He shook like a mouse in a skillet,
So timid and nervous did feel.

We passed by some steamers lights blazing,
And Maurice he whispered to me,
"Ain't that a fine sight Mr. Daly?
Apothecaries' shops on the sea!"

"Does a swordfish go in for hop bitters?"
Says Maurice to me with a frown,
"Is there no one laid up with the measles?
For it's strange to see drug stores leave town!"

We struck the white-coats the next morning,
And over 'er side every man,
With his gaff and his bat on his shoulder,
As we copied over each pan.

And Maurice a half mile behind us,
Was catching all kinds of queer frills,
He was bowing and scraping on tip-toe,
Like a man in a set of quadrilles.

We missed him all day till the evening,
I mind I had four young harps laced,
We were over three miles from the steamer,
A long road before us to face.

McCarthy had six, he was stronger,
And Mullins could only haul two.
I said to myself, "Oh, yes, Crotty
Was the only slack man in the crew."

Coming home 'bout a mile from the steamer,
We saw Maurice stripped off for a bout,
And a big old do-hood with his flippers,
Was stretching him out every clout.

"I challenged him fair," said poor Maurice,
"For a fight he before me does stand
But he took a mean dirty advantage,
And he hit me with rocks in his hand!"

We backed him in turn to the steamer,
And tucked him up snugly in bed;
Next morning he came to his senses,
He called me asidc and he said:

He must have got drunk from the liquor,
Or with that he would beat me to death,
For I'm certain he had a nice jag on,
I got the smell of 'Old Tom' from his breath.

The Spring of the Wadhams

I think 'twas the spring of the Wadhams,
We were out in a ship called the *Dan*,
Maurice Crotty was one of our seamen,
A comical, queer, funny man.

He'd sing songs and tell stories forever,
And lies he would spin by the score,
And when Maurice would kick up a ditty,
All hands in the galley would roar.

It was his first spring out ice-hunting,
Not a rope in the ship did he know,
He didn't know the first thing about her,
Not even to lace up a tow.

The captain came on deck one fine morning,
Says he, Maurice, your trick at the wheel;
He shivered like a bag of live kittens,
He shivered from mizzen to keel.

The captain went into the cabin,
To have a quiet nap and a smoke,
And he rushed out on deck in a hurry,
And fierce, like a tiger, he spoke!

"What course are ye steerin', ye booby,
You jackass, you blunderhead goose."
If she keeps on her course, sir, say Maurice,
We'll strike on the back of Renews.

What course are ye steering, cried the captain,
The vessel being down on her side;
Oh! says Maurice, if the swell doesn't trip
her,
We'll strike on the back of Kilbride.

I leaned o'er the rail one fine morning,
And I saw poor Maurice fall in,
And I reached out a gaff for to haul him out,
For the water was up to his chin.

Reprinted from John Burke and George Oliver, THE PEOPLE'S SONGSTER, BUYERS' GUIDE AND GEMS OF POETRY AND PROSE. (St. John's: Burke and Oliver, 1900), 12.

I Am a Newfoundlander

I am a Newfoundlander, I go out to the ice
I'm always in the best of ship, I take it
for a choice
I first look at the captain to see his
northern jaw
The man I wish to sail with is Captain
Harry Dawe.

In 1906, my boys, to the country first she
came
And in a few days after, she proudly spread
her name
They called her the *Adventure* and set her
on a trial
And 20,000 whitecoats were captured in their
prime.

The ship that I do speak of was the last to
get up steam
She is a smart and speedy boat that can't be
overhauled
No matter where the seas may be if they are on
the flow
Where Captain Dawe does point her she's
certainly sure to go.

The morning of the tenth of March the town was
merry and gay
To see each ship hoist up her flag and proudly
steam away
In every ship 200 men under their captains
command
And every man a credit to his country Newfoundland.

The ship that I do speak of was the last to get
up steam
And when the others they had gone she hauled out
in the stream
Then Captain Dawe came on the bridge full speed
When just a half hour after *The Adventure* took the
lead.

Each man felt bright and happy as she did
speed along
Some would tell a story, some more would
sing along
Some had a glass of grog along some more had
a kettle of brewis
The free men in their glory as they shoveled
in the fuse.

We had an Irishman on board from Dublin town
he came
We also had a German with an aristocratic name
The Irishman went down below to oil both crank
and wheel
And when he accidentally lost his foot from
instep to the heel.

You'd pity his misfortune as we brought him up
again
The doctor did his very best to relieve him of
his pain
This happened about noon, my boys, about 50
miles from shore
And we landed him with kind hearted friends at
a quarter after four.

After leaving Catalina, we soon do meet the ice
She passes heavy pans aside as if 'twas only rice
The second day out from St. John's, no better yet was done
But we struck the whitecoats in their prime at the setting of the sun.

And now is Monday morning and there's a long week ahead
And in the body of the whitecoats each man is painted red
On Saturday night through a frosty fight we hauled, we pulled and dragged
We had 30,000 stowed below, 10,000 more were flagged.

And now it's Monday morning each man gets gaff and rope
O'er the side like hardy boys with spirit and with hope
And Captain Dawe came on the bridge, he spoke to all his crew
"I want you all to work like men for you've got lots to do."

But harder still were those to kill as we poor men did do
The pain steamed in among our flags and then put out her drew
It's not the crew that we do blame or about them we do shout
If the captain hadn't brought them they could not get out
It's the captain that we do blame, not the hardy crew
For they do want a whitecoat as well as me or you.

This is a song which originated around the seal fishery of the early 1900s. It was recalled to me by my grandfather and sung to the tune of "The Star of Logy Bay."
(Eileen Colbert)

The song was collected for a MUNFLA project (card 70-11) by Eileen Colbert of St. John's from her grandfather, Mr. Frederick Dodd. Printed here with permission of Mrs. Michael Colbert and Dr. Herbert Halpert.

A Sealer's Love Letter

Dear Miss: — I know I can't mail this
Forgive me, it's all I can do,
Out here at the ice-fields in Winter
So I'm writing this letter for you.
For it's Easter good wishes I'm sending
Though in words not classic or nice
To you where South River is wending
From me out here at the Ice.

Perhaps you've forgotten the morning
When the train for St. John's pulled away,
I parted from you without warning
And signed as a sealer next day.
I was working last Winter in home town
And I then got acquainted with you
The pleasure, and promises we made dear,
And...well, I remember, do you?

You were there for a short while, at least, Miss
Then the whistle, a word, a last smile
And I went away on the train, Miss
And you went away to your toil.
You, home to your world and its beauties,
Your dances, your parties, and shows;
I, back to a sealer's rough duties,
In sunshine, and rainstorm and snows.

But to-night I'm alone on my watch here
By the quarter deck's rattling wheel chain
While dog hoods are yelping out back here,
You'd be scared, Miss, I guess by the same.
The moonlight is white on the ocean
And the long frozen miles of the ice
Seem to shrink in the North wind's cold motion,
But I still think it's awfully nice.

It's different where you are, I reckon
Leastways, I think it must be
Where the green fields of Newfoundland beckon
And South River swings down to the sea.
You like it, no doubt, you've no notion
Of the adventure that sealing trips hold
Where the snow lies in drifts on the ocean
And the ice-fields are empty and cold.

But somehow I wouldn't forego it
For all of the pleasure at home.
I breathe it, I feel it, I know it
And it thrills me clear through to the bone
The stars in the night, how they glisten,
The ice in the day, how it's spread;
There's room to stand up and to listen
And know there's a God overhead.

And then when the Springtime is coming
And white-coats to the water will take,
You hearken at dawn to the splashing
Of old seals down in the wake.
Why, home town simply ain't in it;
But, Miss, am I writing too free?
Excuse me my thoughts for a minute
Strayed from the *Terra Nova* and me.

As it's just about Easter, my notion
Was to send you good wishes and such,
From the trails that lie over the ocean,
So, I guess it don't matter so much.
If you'll forgive me an uncultured party
In the spirit of Easter, I'll thank
And send you good wishes all hearty
From yours most sincerely, Frank.

Reprinted with permission from THE FAMILY FIRESIDE. St. John's: Gerald S. Doyle Ltd., n.d.

Loss of the S.S. Algerine

Attention all ye sailor boys
And hark to what I say
And hear about the *Algerine*
Was lost in Hudson Bay,
One of the old time sealing fleet
As well we all do know,
That often sailed the tenth of March
To face the Northern floe.

The *Algerine* some months ago
Did boldly sail away,
With parties from the stars and stripes,
To explore Hudson Bay.
Beside the strangers on the ship
That went down to explore,
Her crew all lads from Newfoundland
And natives of the shore.

She met much ice upon the voyage
And stormy every day,
Until the *Algerine* did reach
The shore of Baffins Bay,
With icebergs dotted all around,
A splendid sight to view,
And something they'll remember
The explorers and the crew.

When in the heavy rafting ice
That's packed near Baffins Bay,
The *Algerine* got tightly squeezed,
Her timbers soon gave way,
Her crew were forced to leave the ship
The strain she could not stand,
Provisions, punts and lifeboats,
For to try and reach the land.

They reached the land in safety,
And for weeks they had to stay,
And anxiously stood watching
For a steamer every day,
And with provisions very scarce
To feed the steamer's crew,
Perhaps forced to stay the winter
Made the scene look awful blue.

At last their spirits did revive,
From many an anxious night,
And soon they saw a steamer's smoke,
The *Neptune* hove in sight,
And Lucky Scott and his brave crew
Lost very little time,
To save the shipwrecked sailor boys
From Bowring's *Algerine*.

The Algerine was lost in 1912 in Hudson bay while on an American Expedition. (Eds.)

Reprinted from Johnny Burke, BURKE'S BALLADS. (St. John's: Johnny Burke, 1912), 24.

In Memorial of 77 Brave Newfoundland Sealers

Sad comes the news from oe'r the sea
To fill our hearts with dread,
To tell us that the ones we loved
Are numbered with the dead.
A husband dear a loving son,
Or one we did adore,
Died on the bleak and northern floe
Alas, we'll see no more.

They left their homes with happy hearts
In health and spirits gay,
And bid their friends a fond adieu
As they sailed out that day;
The pride and beauty of our land
Who knew no dread or fear,
None thinking as they left their homes
That death was drawing near.

May God who does things for the best
We pray to give relief,
To mothers, wives and children
To bear this heavy grief;
And may they see a brighter land
Most fervently we pray;
For these brave souls who lost their lives
Upon that fatal day.

Who lost Their Lives in a Blizzard at the ice March 31st, 1914.

This Broadside is in all probability the work of Johnny Burke. It was copied from an original now in the possession of Mr. Herbert Cranford of St. John's. Courtesy of Paul Mercer.

The "Newfoundland" Disaster (1)

We often read of heroes bold and noble deeds they done,
Some on the field of battle Victoria Crosses won,
The British tars and officers who walked the quarter-deck
Oft la'nched a life-boat in a storm to take men from a wreck.

The hero that we speak about his praises for to tell
Is plucky Captain Randall commander of the *Bill*,
Who put his ship about that day to rescue he did go
And saved the lives of twenty-five out on the northern floe.

For two long days and dreary nights these poor lads had to stand
Caught in a blinding snow-storm, the crew of the *Newfoundland*,
Fatigued with hunger, thirst, and cold and no relief in sight,
When seventy-seven brave lads succumbed upon thc ice that night.

On Thursday morn with tottering steps the few that did remain
Made for the *Bill* as best they could, the ship did try to gain,
And Captain Randall from the bridge these poor lads he did sight
Who told their sad and dismal tale how for their lives did fight.

The captain then gave up the voyage, for St. John's he did steer,
To bring the dead and dying to the ones they loved most dear,
The kindness they received on board no human tongue could tell,
Of the kind-hearted sailors and the captain of the *Bill*.

In after years his name will live and hang on memory's wall
To show their children's children the heroes of them all.
And his kind-hearted sailors their great and daring band
Who saved the sole survivors of the steamship *Newfoundland*.

The Newfoundland Disaster (2)

Come all ye sons of Newfoundland
And shed a tear or two
While I relate the hardships great
Befell this steamship's crew
Upon the thirty-first of March
That sad and fatal night
The *Newfoundland's* bold hardy crew
Got lost upon the ice.

About one o'clock that evening
A gale came on to blow,
Which afterward was followed,
By blinding drift and snow.
Methinks I can imagine some,
How they sat down and wept
While others fought until the end
With the Grim Reaper Death.

It must have been a ghastly sight
To see him going past,
The monster Death upon the ice,
To form his gruesome task
And as the night wore slowly on
The wind with fury blew,
His icy hand was laid on some
Who formed that galland crew.

The strong men bowed beneath his
grasp
The youthful head bent low
To Him, who rolls the storm along,
And cheers the weary soul,
When dawning broke on Thursday
morn
The storm it did abate,
And nearly eighty sons of toil
Upon the floe lay dead.

Oh Newfoundland! Oh Newfoundland,
Who mourn your noble dead
Who often trod the frozen pans
And from it wrung their bread;
Now in this hour of deep, deep gloom
Combine and do your best
To ease the ones now left to mourn
The wife and fatherless.

No more their footsteps shall be heard
Those true and faithful dead
No more we'll gaze upon their face
As oftimes 'fore we did.
Their names will live in history
And hang on memory's wall,
As true born British heroes,
Who died at duty's call.

(Geo. Humbey, Summerville, B. Bay.)

Reprinted from THE HARBOUR GRACE STANDARD.
Harbour Grace, 1914.

Born in 1883 one of the oldest of the old sealers goes about his duties as returning officer in Riverhead, Harbour Grace during the 1975 Provincial election. Edward Russell sailed with Captain Bob Bartlett in 1913. His brother, Jack, was lost on the ***Southern Cross*** *in 1914.*

Photo by Cyril Bishop.

The Terra Nova

One Monday morning March the Tenth,
It opened fine and clear,
The slob ice in the Narrows
From St. John's to Cape Spear.

But nothing daunted Captain Kean,
Upon the bridge he stood,
In the good ship *Terra Nova*,
To search for harp or hood.

Three blasts came from her whistle
Which woke up all the town,
And men from all directions
To wharf came running down.

Her roll was called, all hands aboard,
A hearty lot of men,
Bound to the northern ice floes
To fill her up again.

Another blast and then she hauled
Away from Harvey's Pier,
Her course out through the Narrows
The helmsman straight did steer.

She hugged her way to the narrow head
And soon was put about,
But the slob ice was getting hard
It was without a doubt.

With staysails set to catch the breeze
She forced ahead full steam,
In half an hour the captain said,
We'll have Torbay abeam.

But very shortly after
A pan got in her way,
Which caused the death of three bay boys
Way out in St. John's Bay.

A dozen of the pluckiest jumped on the ice that day
To help the *Terra Nova* to carry on her way,
But very shortly after, sad news it flashed to town,
That three of our brave fellows in sight of all went
down.

There was David Wheeler of St. John's,
With a wife and children four,
Who was left without their bread winners
It would make your heart feel sore.

There's Hubert Hiscock of Trinity Bay,
All taken in his prime,
His birth place sent out heroes
To every land and clime.

And lastly comes young Mercy Breen
Down from the Southern Shore,
Where the hardest boys are born and bred,
They've proved it o'er and o'er.

May the God of love who reigns above
Console those friends we pray
Of those brave boys who gave their lives,
Out in St. John's Bay.

This song was collected in September, 1967 by Dr. Herbert Halpert and Samuel Fiander Jr. from Mr. Norman Payne of Ramea, Newfoundland. Printed here with permission of the singer, Mr. Norman Payne and Dr. Herbert Halpert. (MUNFLA Transcript C 445).

Southern Cross

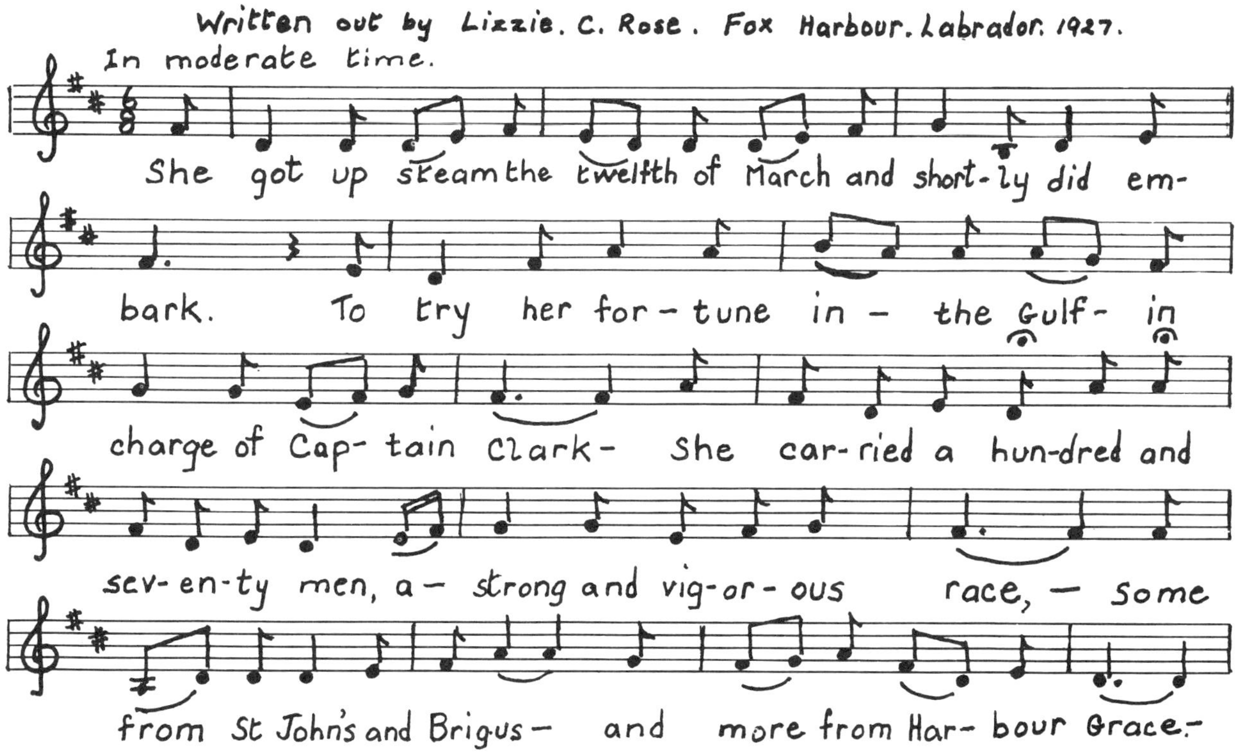

She got up steam the twelfth of March and shortly
did embark.
To try her fortune in the Gulf in charge of
Captain Clark.
She carried a hundred and seventy men,
a strong and vigorous race.
Some from St. John's and Brigus
and more from Harbour Grace.

She reached the Gulf in early March, the white-
coats for to slew,
When seventeen thousand prime young harps
killed by her hardy crew,
All panned and safely stowed below, with colours
waving gay,
The *Southern Cross* she leaved the ice, bound
up for home that day.

She passed near Channel homeward bound, as
news came out next day,
To say a steamer from the Gulf she now is on
her way.
"No doubt it is the *Southern Cross*," the operator
said,
"And looking to have a bumper trip, and well
down by the head."

The last of March the storm came on with
blinding snow and sleet;
The *Portia*, bound for western ports, the
Southern Cross did meet;
When Captain Connors from the bridge he
saw the ship that day,
And thinking she would shelter up in
St. Mary's Bay.

St. Mary's Bay she never reached, as news came
out next morn.
She must have been all night at sea, out in
that dreadful storm.
No word came from the *Southern Cross* now
twenty days or more;
To say she reached a harbour around the
western shore.

The *S.S. Kyle* was soon dispatched to search
the ocean round,
But no sign of the missing ship could anywhere
be found.
She searched Cape Race and every place until
she reached Cape Pine,
But of the ship or wrecking the captain saw
no sign.

The *Southern Cross* out twenty days, she now
is overdue;
We hope, please God, she'll soon arrive and
all her hearty crew,
But put your trust in Providence and trust to
Him on high
To send the *Southern Cross* safe home and fill
sad hearts with joy.

All things do happen for the best, but if they're called
away,
The brave lads on the *Southern Cross*
out in the storm that day,

We trust they reach that heavenly land
and rest with Him on high,
Where cares and sorrows are no more,
but all is peace and joy.

Written out by Lizzie C. Rose, Fox Harbour, Labrador, 1927. (Original note)

Capt. Frederick Harris and the Grates Cove Seal Killers of 1915

Attention all, both great and small,
 A tale I have to tell
Of Captain Frederick Harris
 And young *Florizel.*
He goes out with the Stansford crew
 The pluckiest on our shore;
He takes the lead of Grates Cove fleet,
 Our hardiest Commodore.

The next is William Henry's sons,
 Face every breeze they would
Most every day they land on shore
 With a harp or a dog hood.
The Norris boys are never behind,
 Who braves the Hart Cove gales,
And hauls up fifty quintels of fish,
 With their prime bedlamer seals.

There's Simeon Benson, I may say,
 Has travelled around our shores,
Through heaps of snow and storms of wind,
 He've brought ducks in galore;
Also for foxes and for seals,
 For birds and codfish, too,
He is the gunner every spring
 Of Thomas Avery's crew.

There's William Stansford on the route
 Before the break of day,
And on the way for Bugdie's Head,
 And captures five one day.
While Henry Meadus for a shot,
 No better on our shore,
He captures ten, two days of them,
 Here's Johnny with five more.

Oh, here comes the Sam Duggan boys,
 White jumpers, they look red,
With a fine tow on breakfast hour
 I killed, brave Samuel said.

Some more fine lads from Daniel's Cove,
 Did travel the ice-floe,
And captured a dog harp that day,
 Took two of them to tow.

'Twas on the 30th of April last,
 Captain Harris leaves our shore,
And by the hour of 11 a.m.
 E. Stansford captures four.
They were then close by old Baccalieu,
 With famous Andrew Hatch,
Who captured five, while just inside,
 Skipper Thomas strikes the patch.

Most of the fat was seen last spring,
 Quite near old Baccalieu.
Captain Harris pulled on the aft oar,
 And ordered his fine crew
To keep out with the Eastward fleet
 When his sealing days are done,
And all we ask at his depart,
 To leave us ship and gun.

Success to Captain Harris,
 And to his crew as well,
May he have as good brave men next spring,
 In the young *Florizel.*
So now good luck Bill Winsor,
 Bill Bartlett you the same;
There's a cry coming from the East and West,
 Oh, now where's Bumper Kean!

By Joshua Stansford, Grates Cove.

Reprinted with permission from Joshua Stansford, FIFTY YEARS OF MY LIFE IN NEWFOUNDLAND. (Devon: A.H. Stockwell, 1952), 70.

Au Revoir To Our Hardy Sealers

Our gallant ships are going, where rude Boreas is a blowing,
With a bitterness that blanches many a stout and rugged form,
But we pray each hardy "swiler," every honest, sturdy toiler,
May come back safe and solid from the ravage of the storm.

Oh, farewell, and may God bless you, in this language we address you,
You, the sons of valiant seamen, who had sailed in days of yore,
May kind Heaven hover o'er you; may her smiles go on before you,
Terra Nova's sons and daughters, truly bid you au revoir.

Where the icebergs roll and rumble, where the salty billows tumble,
Where, with terrors of old Neptune, you're beset on every hand,
There is not, the wild world over, any other salt sea rover
Who combats with such dangers as the men of Newfoundland.

When the seas are roaring frantic, on the blue and broad Atlantic,
Then you're found upon its bosom, on the bleak cold Labrador,
And in spring you go a sailing with a courage that's unfailing,
To the icefields, may God bless you, as we wish you au revoir.

Reprinted with permission from James Murphy, SONGS OF NEWFOUNDLAND BY VARIOUS AUTHORS. (St. John's: James Murphy, 1917), 8.

The "Kite" Abandoned in White Bay

(22 Men Leave Her.)

Come all ye rambling sailor boys
  And hearken please to me,
And hear what fishermen endure
  Who plough the raging sea;
For at the ice, as well ye know,
  The hardship they must stand.
Exposed to bitter wind and cold,
  The boys of Newfoundland.

The *Kite*, she sailed the 12th of March
  Bound to the northern floe,
And not as powerful as the rest
  As all of you well know;
For she was easily left behind,
  While others took the lead,
The *Kite*, was forced to lag along
  She hasn't got the speed.

For she got jammed in White Bay
  And there she lay caught fast,
While the main patch of the seals
  She accidently passed.
She still remains caught in the floe
  With scarcely any seals,
And while her crew are in despair
  Their said misfortune feels.

Some twenty-two of her brave crew
  Struck out and made the land,
Perhaps a summer's work to lose,
  The thoughts they could not stand.
And five were boys from St. John's town
  Who landed here to-day,
And more were most outharbour men
  Belong to round the bay.

Reprinted from an unidentified manuscript presented to MUNFLA by Mr. Leo Moakler (n.p., n.d.). In all probability the work is that of Johnny Burke. Courtesy of Paul Mercer.

Andy Short, born in 1900, sailed on a number of famous vessels including the ***Viking****, the* ***Thetis****, and the* ***Eagle****. He was also with Abram Kean when that famous Captain reached his goal of 1,000,000 seals and he was aboard the* ***Ungava*** *when Peter Carter brought in what was probably the heaviest load of seals in one trip in 1933. Andy also qualified and worked as a Master Watch. The former Sealing port of Harbour Grace is in the background.*

Photo by Shannon Ryan.

To the Memory of the Late Captain Kennedy

Slowly to-day we wended our way
  To a grave in Belvedere,
Behind the corpse of a hero bold,
  By all his name held dear.
He died that others might live,
  No greater love hath man
Then he lay down his life for a
  friend,
  This seaman of our clan.

For many a year he knew no fear,
  On shipboard he took delight,
And with his hardy shipmates
  Spent many a weary night.
Yet Providence protected him
  And brought him safely to shore;
But alas! his life is ended
  And he'll strike his course no
  more.

On board the *S.S. Viking*
  To the sealfishery he set sail,
As navigating officer
  His profession did entail.
This hero son of Newfoundland,
  This Captain of the sea,
Was never known to shirk his post,
  As shown so recently.

The scene of that disaster
  On that dark and dreary night,
In fancy I can see it,
  I can ne'er shut out the sight —
The lurch, the shock, the explosion,
  The cry of "ship on fire."
The frantic cries of his wounded
  mates
  In that distress was dire.

Those cries were not unheeded
  By the hero of this song,
Though badly bruised and bleeding
  He fought his way along
To where his shipmate, Clayton
  King,
  A hero true and brave,
And by a superhuman effort
  Saved him from a watery grave.

We can see another suffer,
  Sargent — American by birth,
A passenger on the *Viking*,
  Who also proved his worth;
This hardy son of Uncle Sam,
  A hero staunch and brave,
By kind Providence protected
  From a North Atlantic grave.

For forty-eight long hours
  They fought desperately for life,
Clinging to that piece of wreckage
  In the midst of death and strife,
But spurred on by the courage
  Of this brave man of the sea,
Young Clayton King was rescued,
  His friends once more to see.

This is just a brief synopsis
  Of pluck, daring and of grit
In such an awful circumstance
  of a man who did his bit.
May his family all be comforted
  By his noble work well done,
For in making the supreme sacrifice
  He has reached that heavenly
  throne.

By an Unknown Author

Reprinted with permission from Clayton L. King, THE VIKINGS LAST CRUISE. (St. John's: Evening Telegram Printers, 1936)

The Swiles of Newfoundland

"Oh, the day we left St. John's, me b'ys,
It was a very fine day!
Our wives an' sweethearts on the quay
Says they, ye'll understand,
Farewell, says they, remember me
To the swiles o' Newfoundland!

Our duff was hard, our pork was bad,
We had to drink pinnacle tea!
Hurrah, me b'ys, we'll make a noise,
An' by our gaffs we'll stand!
Me rope upon me shoulder,
Me gaff all in me hand,
Both day an' night 'tis my delight
To kill swiles in Newfoundland"

Reprinted with permission of Tundra Books of Montreal from George Allan England's THE GREATEST HUNT IN THE WORLD. (1969), 128. Originally issued in 1924 under the title VIKINGS OF THE ICE.

Uncle Bill Teller

"Uncle Bill Teller died las' fall,
Young maiden, where ye bound to?
We jigged t'ree days an' niver got one,
Acrost de Western Ocean.

Oh, Billy K. is de divil fer fat,
Hang to 'er, b'ys, hang to 'er.
He can allus smell where de swiles is at,
Hang to 'er, b'ys, hang to 'er!

Billy K. got a fine old bark,
Hang to 'er, b'ys, hang to 'er.
He'll strike de patch afore it's dark,
Hang to 'er, b'ys, hang to 'er!"

Reprinted with permission of Tundra Books of Montreal from George Allan England's THE GREATEST HUNT IN THE WORLD (1969), 129. Originally issued in 1924 under the title VIKINGS OF THE ICE.

Willy March

"De home of his childhood, in Nothren Bay,
He quit it fer pleasure, much more 'an fer pay.
On de icefields he ventured, most yout'ful an' brave,
Whereon he fought death, but his life could not save.

Dat cold night in March he lay miles from de shore,
From friends dat was watchin' to see him once more.
But alas, human aid may be looked fer in vain,
Fer life was too sweet on de ice to remain.

At dawn, Easter Saturday, all happy they would be
If in de Bay of Conception a sail dey could see,
But no sound from de ocean could bid dem rej'ice,
Or lead dem to think he was safe on de ice.

Wid courage undaunted he set out anew,
To reach Cape St. Francis, de object in view;
When three of his comrades whose number was four,
Lay on deir cold beds, not to walk wid no more.

He an' de odder set out fer de Cape.
How hard it must dwell on dat poor feller's fate!
He found himself weak, as he got near de shore,
An' he said to de odder, I cannot walk more.

He said to de odder, Go on fer de Light.
Tell dem at de house to come fer me to-night!
Dis man reach de lighthouse, which bein' quite nigh,
But little he t'ought he layve Willy to die!

Oh, rugged Cape St. Francis! re-echoed de v'ice
From him who fer aid called loud on de ice.
He expected assistance, but none did resayve,
An' he died near de shore, where de lost might be
saved!"

Reprinted with permission of Tundra Books of Montreal from George Allan England's THE GREATEST HUNT IN THE WORLD (1969), 126. Originally issued in 1924 under the title VIKINGS OF THE ICE.

The Sealers of Newfoundland

Ho! We be the Sealers of Newfoundland!
We clear from a snowy shore,
Out into the gale with our steam and sail,
Where tempest and tumult roar.
We battle the floe as we northward go,
North, from a frozen strand!
Through lead, through bay, we fight our way,
We Sealers of Newfoundland!

Yea, we be the Sealers of Newfoundland!
We laugh at the blinding dark;
We mock the wind as we fling behind
The wilderness hoar and stark.
We jest at death, at the icy breath
Of the Pole, by the north lights spanned.
In a wild death dance we dice with chance,
We Sealers of Newfoundland!

Sealers, ho! Sealers of Newfoundland,
With engines begrimed and racked,
With groaning beams where the blue ice gleams,
We push through the growlers packed.
With rifle, with knife we press our strife;
What lubber shall understand
The war we fight in the ghostly night?
Aye, Sealers of Newfoundland!

The ice glows red where our skin boots tread,
And crimson the grinding floes.
From mast we scun till our race be run,
Where the Labrador current goes.
From ship we spring to the pans that swing;
By stalwarts our deck is manned.
O'er the blood-red road the sculps are towed
By the Sealers of Newfoundland.

Oh, some may sail with a southern gale;
Some may fare east or west.
The North is ours, where the white storm lowers,
Wild North that we love the best!
O North, we ken that ye make us men;
Your glory our eyes have scanned.
Hard men, we be, of the Frozen Sea,
We Sealers of Newfoundland!

Bitterly bold through the stinging cold
We vanquish the naked North.
We make our kill with an iron will,
Where the great white frost stalks forth.
"Onward!" we cry, where the bare bergs lie,
Dauntless our course is planned.
With blood, with sweat, scant bread we get,
We Sealers of Newfoundland!

"Starb'rd!" and "Steady!" and "Port!" we steer;
Press on through the grinding pan!
We labour and muck for a fling at luck,
Each man of us, God! a man!
We cheer at the bawl of the whitecoats all,
We labour with knife and hand,
With rope and gaff. At the North we laugh,
We Sealers of Newfoundland!

Where the old dog hood and the old harps' brood
Lie out on the raftered pack,
We tally our prey. Then away and away,
Men, ho! for the homeward track!
Till the day dawns near when a rousing cheer
Shall greet us, as red we stand
On the decks that come to our Island home,
We Sealers of Newfoundland!

Written by George Allan England

Success to Every Man

"De time is drawin' near, me b'ys,
De narthern floe to face,
So we must get out 'aulin' rope,
De whitecoats fer to lace!
Our belt, skin boots an' shaythin' knife,
Our piper an' our pan,
An' rough it ah-h-h-h-ahl de month o' March,
An' do de best we can!

Some in de Gulf is goin' to sail,
De whitecoats fer to fight,
De *Viking* an' de *Suthren Cross*,
De *Nipshun* an' de *Kite*.
De warriors o' de wooden fleet,
Dey soon will sail away,
In charge of 'ardy swilin' crews,
Wid colours flyin' gay!

A few more days, an' ahl will sail,
Doze steamboats strong an' good,
To face de harp an' bellamer,
Whitecoats an' ole dog hood,
An' soon de slaughter will begin
Upon de bloodstained pan,
An' may good luck attend de fleet!
Success to every man!

On Patrick's Day, wid colours gay,
When in de narthern floe,
We hopes ye'll have your steamer full,
An' safely stowed below,
An' ahl return wid heavy loads,
An' buntin' wavin' gay!
Wid cheers from doze as is on shore,
When you return, some day!

Den, yere's success to ahl de fleet,
De wood an' ironclads,
Manned by de yout's of Newf'un'land,
Ahl active, hardy lads!
Fer ahl can do deir part like men,
An' face de frost an' sleet,
To haul a tow in rain or snow,
Our b'ys dey can't be beat!

May Providence guide ye, night an' day,
An' *suc*-cess crown y'r work!
Fer when de swiles is dere to kill,
From work ye niver shirk.
An' may ye ahl make tidy bills,
An' wealth to every man!
Fer he deserves a-a-a-ahl he can make
Who treads de frozen pan!"

Sung by Tom Fillyards

Reprinted with permission of Tundra Books of Montreal from George Allan England's THE GREATEST HUNT IN THE WORLD (1969), 233. Originally issued in 1924 under the title VIKINGS OF THE ICE.

The Bird Rocks

In moderate time.

'Twas winter down the icy gulf,
The Gulf St. Lawrence wide,
Where stands a lighthouse on a rock,
The sailors' friend and guide.

The keeper had his wife and son,
A helper, too, had he;
These four alone lived on that rock,
Surrounded by the sea.

One day these three brave men went out,
As they did want to do,
On rugged sheets of frozen ice
To capture seals a few.

But as they lingered o'er the swile,
At length they failed to see
The wind had veered from south to east
And drove the ice to sea

The sheets of ice on which they stood
Were drifted all unseen;
Now they could not gain that rock,
For water flowed between.

Alone within the lighthouse tower
The keeper's wife did see
Her loved ones carried to their doom
Toward the open sea.

They wove their hands in wild despair
And did for rescue pray,
But not a soul was near to help
That freezing winter's day.

Four months she lived on that lone rock,
Grief-stricken, yet so brave,
For nightly gleamed her beacon light
Across the gloomy wave.

At length the welcome spring appeared,
The steamship came around,
But when the skipper climbed the rock
But one brave soul he found.

She told her tale in plaintive tonc;
The skipper's eyes grew dim;
He turned his eyes towards the sea;
His heart was touched within.

"What did you do?" at last he said,
"When all was snatched from you?"
"I kept my light still burning, sir;
'Twas all that I could do."

O, may we, like this faithful soul,
In sorrow's darkest night
Still do our duty for our God
And show the world our light.

Then let us keep our lights aglow,
Till all life's storm is o'er;
Then we'll shine as God's fair jewels
On Eden's peaceful shore.

Sung by Mrs. Annie Walters, Rocky Harbour, 1929.
(Original note)

The sealing fleet passes through the narrows of St. John's in 1902 on its way to the ice fields.

Courtesy of Baine Johnston & Co., Ltd.

The Cooks of Torbay

Come all ye young fellows wherever ye be,
I'll sing ye a verse on the cooks of Torbay,
And if ye'll pay attention and listen awhile,
You'll hear a tall song that will cause you to smile.

We signed on at the office a-sealing to go,
Up in the Gulf, in the *Ellen*, you know;
Our cook he lookéd drowsy, those words did he say:
"O, you'll find me smart enough to sea."

We hove up in the narrows, and then rowed ashore,
'Twas down to Malone's he stayed, stretched on the floor.
. . . .
. . . .

Our skipper come on board, these words he did say:
"Well, Bob-din, my boy, there's no dinner to-day.
By great help you've cooked us two meals, and boiled us no pork;
When it comes to the third one, the hell he bobbed all."

He went down in the fo'castle and he begs for to grieve;
He says to himself, "I'm enchanted, I believe,"
But rising again said, "I'm one man in ten;
So Tabeau, you must leave the galley again."

He went down to the cabin and began for to grunt,
Shoved up his three fingers and likewise the stump;
He says to his shipmates, "I won't cook at all,
For the coppers is rusty and the galley too small."

Now Tabeau mixed his duff well, you do know;
He have his young son for to tend on the crew;
He'll sleep in his park until all is complate,
And then he'll go forward and give the men prate.

Sung by James Day, Fortune Harbour, 1929.
(Original note)

Reprinted with permission from Elizabeth Greenleaf and Grace Mansfield, BALLADS AND SEA SONGS OF NEWFOUNDLAND. (Hatboro, Pennsylvania: Folklore Associates, 1968), 303.

A Noble Fleet of Sealers

There's a noble fleet of sealers,
Being fitted for the "ice."
They'll take a chance again this year
tho' fat's gone down in price.
And the owners will supply them
as in the days of old,
For in Newfoundland the Sealing Voyage
means something more than gold.

The *Algerine* is first to sail
She's steaming out the Harbour,
With eager sealers on her deck,
And on the bridge — Wilf Barbour.
The *Viking* blood runs in his veins,
As in the days of yore
When the Barbours fought the seal and whale
And fished the Labrador.

Chorus

The *Terra Nova's* next to sail
In charge of Charlie Kean,
In the history of our fisheries
That's a grand and worthy name.
His crew of bully, northern men
Can handle gaff or gun,
To get their share, they'll risk and dare,
And think it all great fun.

Chorus

The *Arctic Sealer's* late to sail
Her crew worked with a will,
Led by that modern Jowler
The Sealer's friend — Sid Hill.
Tho' the last to leave the Harbour,
He was first to strike the patches,
And on March the twenty-ninth
Bore-up log loaded to the hatches.

Chorus

There's one sailed from Catalina,
Her owner is Commander,
She's the staunch and sturdy, local built
The good ship *Newfoundland.*
When the whitecoats bawl, he'll risk his all,
Despite hard luck before,
For there's ne'er a man in Newfoundland,
The likes of John Blackmore.

Chorus

And now they're back in old St. John's,
A-sharing out the Flippers,
Let's wish good luck to sealers all
Likewise their gallant skippers.
Tho' Newfoundland is changing fast
Some things we must not lose,
May we always have our Flipper pie,
And Codfish for our brewis.

The Little Swiler

He was such a very little chap
Blue eyes and sunny smile
Sneaked off that morn unknown to Pa
To go in search of swile.

Two miles he trudged, o'er lumpy ice
And the goin' was so rough
He had a sheath knife, but no gaff
Thought he was really tough.

For Pa was sick, he couldn't go
But he, most growed a man,
Could tow a swile, yes that he could
As rough ice he did scan.

He spied some men, there killin' swile
A little way to east,
With hope revived, he hastened there
He'd get a swile at last.

Hullo there son, what doin' here?
They asked him as he came,
I wants to get a swile like yours
Is it whitecoat, he'd exclaim.

They killed a swile and laced him up
He was so proud you know.
When all hands started out to haul
He was 'mongst them with his tow.

Before he had gone very far
His little face grew pale
He sank exhausted there on ice,
He was too weak to haul.

"When did ye eat me son", they said,
Their voice rough with concern,
"Forgot me grub bag sir," said he,
"Had na'ar bite since morn."

They fed him from their meagre store
And he revived again.
He put his towline on his back
And joined the towin' train.

They landed safely in the cove
He leading older men,
For this proud swiler with his tow
Was really only ten.

Printed with permission from the private collection of the late Gerald S. Doyle.

Thoughts on the Newfoundland Sealing Voyage

Heigho, my lads, for the tenth of March,
And a gallant ship and crew;
All spick and span for the northern "jam",
From the keel to the main royal clew.
Then a parting glass to each winsome lass,
As the siren's screech is heard
In measured time along the line
For all hands to get on board.

Brimful of glee, we'll put to sea,
Let the winds blow high or low,
For a foremost place in the white-coat race,
My boys! We must go slow.
When we strike the "jam" each hardy man
Will have his mettle tried,
When he laces in line, five white-coats prime,
And haul them up long side.

The ice will moan, and our ship will groan
From the pressure of steam and sail,
But we'll skillfully "scun" till our race is run,
Nor heed not the North-east gale.
"Get ready a blast to clear her aft,"
With the main yards laid aback
"Full speed," I say, and "fill away"
And open for her a track.

Where the "growlers" pass, and the icebergs crash,
In the home of the old dog-hood.
Thro' the gleaming floe, we will boldly go,
In search of the old harps brood.
And northward, ho! tho' our speed is slow,
While the "scunners" loud to call
"Port" and "Steady" and "Starboard"
"Hark! I hear a white-coat bawl."

Then heigho, my lads, for the tenth of March,
And the wind abaft the beam,
With our sails all set and our whistles wet
And the valves a lift with steam.
Tho' we have to go thro' frost and snow
We are bound to keep a pace,
And fill our ship with a bumper trip
Then heigho for Harbor Grace.

By the late Geo. T. Sheppard, of Cambridge, Mass., March 1926.
(Original note)

Reprinted with permission from THE FAMILY FIRESIDE. St. John's: Gerald S. Doyle Ltd., n.d.

The Master Watch

Three thousand men of the Viking breed
  Will sail for the North today;
An eager throng fills the city streets,
  And loud from the crowded quay
The answering cheer and the shrilly horn
  Proclaim where the ships delay.

But back in a dingy street there lies,
  Enfeebled, and faint, and poor,
Begrizzled, and gray, and weak, a man
  Whose sealing for aye is o'er —
The master-watch of the fearless heart
  Who shall tread the "pans" no more.

Eager, alert and brave was he,
  And fleet of foot as the roe;
Thro' nights of stress he has cheered his mates
  When lost on the running floe,
And his eagle eye was the first to sight
  His ship thro' the whirling snow.

The damp of death is upon his brow,
  And his eyes with mist are dim;
But a passing smile plays for a while
  On the old man's features grim;
He starts as a loud hurrah is borne
  Clear over the eaves to him.

Then wakened memory paints the scenes
  Of many a vanished year;
And he treads the quartered deck again —
  He leads in the lusty cheer —
And lo! he speaks, and his furrowed cheeks
  Are wet with the unchecked tear.

"Ah, yes! I can hear their cheering now,
  They're off to the ice today;
Oh! I've been with them for forty years,
  And know where the white-coats stray;
I know where the main-patch drifting goes,
  Where the seals in thousands lay.

"I've watched the winds and the currents long
  And studied the omens, too,
And none but the veteran sealer knows,
  And none but an instinct true;
I've been in the seals — in my ship alone —
  Alone with a stalwart crew.

"Today they sail, but I may not go,
  Yet memory fondly sings,
'Til in dreams I've found where the harps abound,
  Where the hunter's cheering rings;
My heart is away upon the floe
  Where the blood-stained baton swings.

"I hear the cry of the young all night,
  I long for the morn to shine;
Now I watch three hundred men away
  Over the frozen brine —
I see afar 'gainst the sinking sun
  The flagged-pans lift on the line.

"The crystal berg and the north-lights gleam,
  And the lode star clear o'erhead
Breaks grandly bright on the northern night;
  Our ship once more has led —
And, Captain — Lord, yes! today — "
  Then the master-watch lay dead!

— Dan Carrol.

Reprinted with permission from Gerald S. Doyle, OLD TIME SONGS AND POETRY OF NEWFOUNDLAND. (St. John's: Gerald S. Doyle Ltd., 1940), 77.

"Neptune," Ruler of the Sea

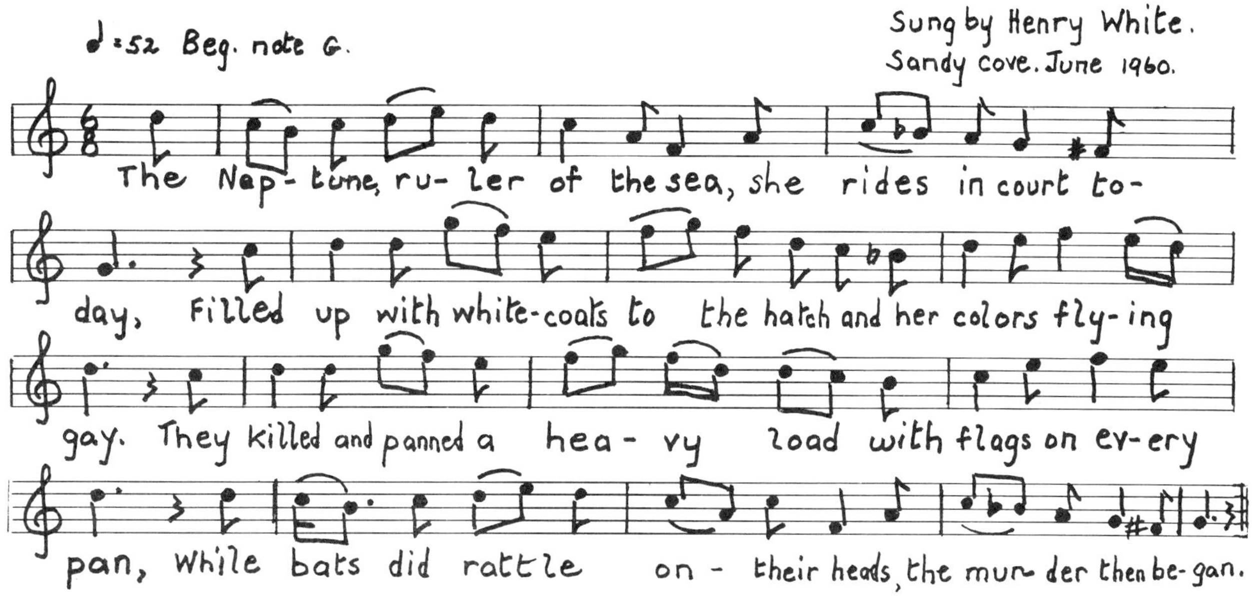

The *Neptune*, ruler of the Sea, she rides in court today,
Filled up with white-coats to the hatch and her colors flying gay.
They killed and panned a heavy load with flags on every pan,
While bats did rattle on their heads, the murder then began.

And not till after the hold was full did they give up the gray,
And then with hatches battened down they bore off for home next day.
'Twas over the rail you could wash your hands; she was so deep in fat,
With thirty thousand prime young harps got slaughtered with their bat.

And here's a truce to Captain Kane and may he live to see
Many springs to fan the fat and fly his long burgee.
And may he live for many a year to load the *Neptune* twice,
Those hardy sons of Newfoundland who trod the frozen ice.

This ballad celebrates a good sealing voyage and proves that not all sea songs are tragic. I have not been able to find this one or the one before it, in print, though both are well established in tradition.

Reproduced by permission of the Minister of Supply and Services Canada from MacEdward Leach, FOLK BALLADS AND SONGS OF THE LOWER LABRADOR COAST. (Ottawa: National Museum, 1965), 208.
Sung by Henry White, Sandy Cove, June 1960
(Original note)

Cotton's Patch (1)

Oh, quite early in March, I remember
the date,
We left for the Ice for the seals to
locate,
We flew round Cape Francis and near
Baccalieu,
But nothing that morning came into my
view.

We flew to the Wadhams and then to
King's Bight
And oh, such a scene, I fell down with
the fright,
I took up the glass and my eye it do
catch,
Some thousands of seals that they call
the Main Patch.

We got in our plane and then home-
ward we flew,
We saw Mr. Bowring and did interview.
"Did the main patch you strike, are
you certain?" says he,
But "mum" was the word from Sid
Bennett and me.

"If you want to find out you must pay
through the nose;
"And we scrammed with the cold and
our fists nearly froze,
"After flying all day over oceans of ice,
"If you want information you must
pay the price."

"Then what is the figure for making
the flight,
"I suppose a few shillings will pay
you all right."
I said, "ten thousand dollars to go
thro' the gap.
"That lets you down easy, you're
getting a snap."

The merchants assembled, the great
Board of Trade,
And ten cents a seal was the price that
they paid,
For every seal that the tally man gave,
And so far we haven't the price of a
shave.

So now, Newfoundlanders, you boast
and you brag
Killing seals, by the fire all day chew-
ing the rag,
All your lives to the ice and you know
every batch,
When I, from Australia, must find the
Main Patch.

Reprinted with permission from John White, BURKE'S BALLADS. (St. John's: privately published, 1960), 14.

Cotton's Patch (2)

We got up steam the ninth of March
And then we sailed away,
And soon was cutting through the gap
For the Northern patch that day;
We left the *Eagle* in our wake,
When rounding Baccalieu
The *Neptune* and the *Thetis*, boys,
And all her hardy crew.

Chorus: —
Oh, didn't we ramble; oh, didn't we
  ramble;
We rambled all around, in and over
  the ground;
Oh, didn't we ramble, scramble,
But the devil a sign of Cotton's patch
  we found.

I was in the barrel all the day
And nearly half the night,
But not a sign of Cotton's patch,
Did ever meet my sight;
At last I said to Wesley Kean,
I think we missed the fat,
May the devil haul you Cotton;
Are you talking through your hat.

I walked nine miles on broken ice,
And frozen to the chin
And like a frozen crackie,
To my ear holes fallin' in;
I turned a stone on Cotton's patch
And said to Mr. Cave,
They are gone on board the *Neptune*,
To George Barbour for a shave.

We then reeled up and put for home,
We didn't strike the batch,
We didn't see a cotton,
Nor half a wincey patch;
The only patch I saw this spring,
And saw it every day,
Was the patch on Tapper's trousers,
He had sewn on in Torbay.

Reprinted with permission from John White, BURKE'S BALLADS. (St. John's: privately published, 1960), 15.

Full Loads to the Sealers

And here's grand success to the sealers,
The pride of our city and town,
Who face the doghood on the ocean,
And with bat like heroes knock down.
Who sail the wild sea for a living
And all sorts of hardship can stand,
The pride of their home Terra Nova,
The brave boys of dear Newfoundland.

In stormy or sunshiny weather,
The sealer must face the wild floe
With a cake of hard tack in his jumper,
And three fathoms for hauling his toe,
And miles on hard ice he must travel,
The home of the white coat to find,
Over fields of rough ice and wild waters,
Laced tidy while hauling behind.

To tread the frozen pans of ice,
The sealer tough and strong,
Well able on a jam of ice,
To haul a tow along,
In frost and sleet thcy can't be beat,
Or on a steamboat's deck,
And very often gets a dip,
In water to the neck.

Success to you sons of bold Neptune,
May good luck attend you this spring
When bound up for home and you loaded,
On the barricade gaily will sing,
With a bill for your wife or sweetheart,
And for your success we all pray,
And the fair maid take for a partner,
The true love you left in the Bay.

Reprinted with permission from John White, BURKE'S BALLADS. (St. John's: privately published, 1960), 28.

The Sealer's Ball

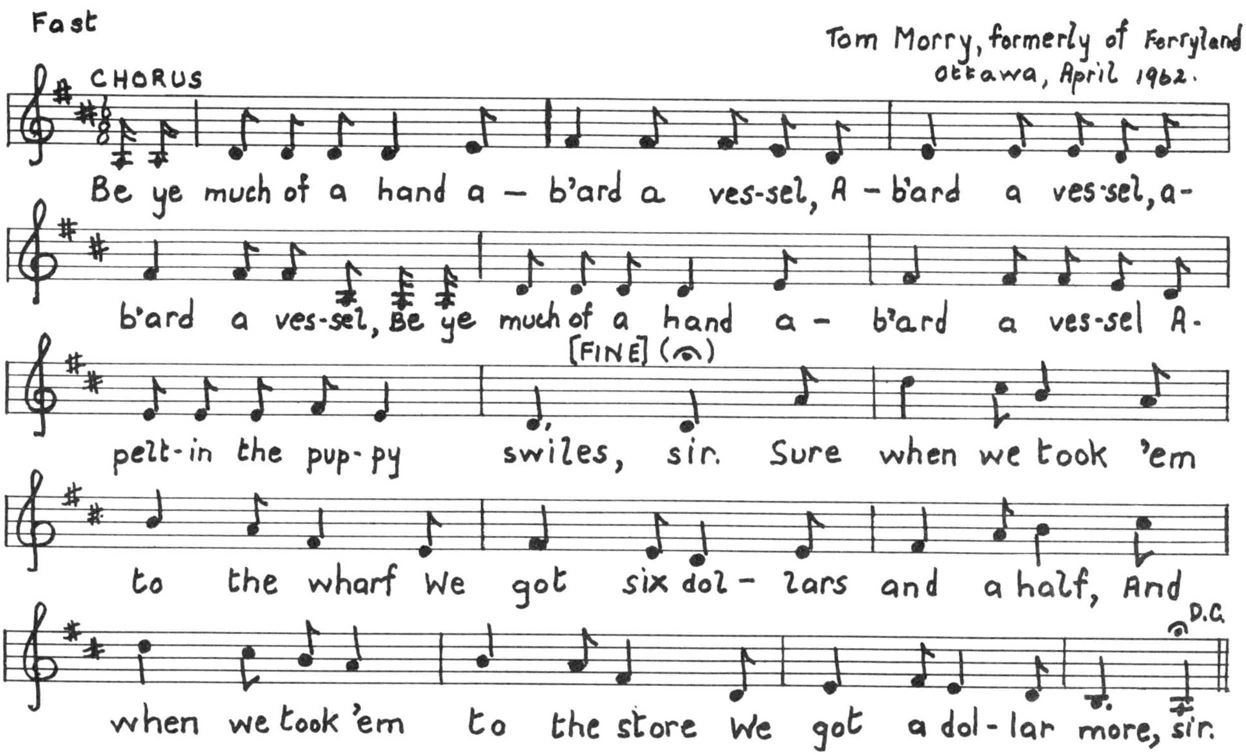

Chorus:
Be ye much of a hand ab'ard a vessel,
Ab'ard a vessel, ab'ard a vessel,
Be ye much of a hand ab'ard a vessel
A-peltin' the puppy swiles, sir.

Sure when we took 'em to the wharf
We got six dollars and a half,
And when we took 'em to the store
We got a dollar more, sir.

We wrote a letter the next day,
And we posted it without delay,
And we sent it off to Jimmy Baird
For a couple of gallon o' rum, sir.

On Saturday evenin' you can see
A couple of gallons they came to we,
We took 'em down without bein' seen
As far as Georgie Wall, sir.

Now Saturday evenin' after tea
A few o' the girls they came to we,
And an elegant time was had by all
Till early in the marning.

A little disturbance then arose
When everyone was pickin' their beaux,
When everyone had picked his own
John Burke he had ne'er a one.

Jack Burke goes up to Jim McGee:
"Now what's ye doin' along wi' she?
She used to go along wi' me,
And she'll do the same this marnin'."

So Jack and Jim get into a clinch,
And ne'er o' them would budge an inch,
And when the clinch broke up
They found the lady she'd a-gone, sir.

Now all young men take lesson o' this,
And never go fightin' about a miss,
'Cause all as you'll do is start a big laugh,
And the lady she'll be gone, sir.

Tom Morry, formerly of Ferryland, Newfoundland.
Ottawa, April, 1962
(Original note)

Mrs. Mullowney Was Three Weeks in Bed Since She Ate the Fipper Stew

Mrs. James Mullowney gave
A party Tuesday night,
And all the aristocracy
Of course she did invite.
They say the table fairly groaned
When Olive laid the spread,
And Mrs. James did honors
To the table at the head.

Old Skipper Jim, her husband,
Sure she worked him like a hoss,
Till his back got in a double
Lugging gads each day across;
Till he nearly filled a puncheon
And to feed a steamer's crew —
For her friends had no fault finding
At this famous fipper stew.

All hands sat in their places,
To do justice to this spread,
And Mrs. Mullowney's plate was fill'd,
You just could see her head;
And to say that fippers vanished
Would be proper to relate,
'Till nothing but a mass of bones
Remained upon her plate.

No doubt she did them justice,
For her appetite was sharp,
She didn't leave a toe-nail
Of a white-coat or a harp.
In fact she gormandized so much
She couldn't eat no more,
Till billious from devouring prog,
She swooned off on the floor.

A doctor soon was hastened,
While she lay in racks of pain,
The doctor said for certain
She had bedlamers on the brain
And now just mention fipper stews
Whenever you go in.
And a lighted lamp of kerosene
Will take you on the chin.

Reprinted with permission from John White, BURKE'S BALLADS. (St. John's: privately published, 1960), 49.

The Sealing Cruise of the Lone Flier

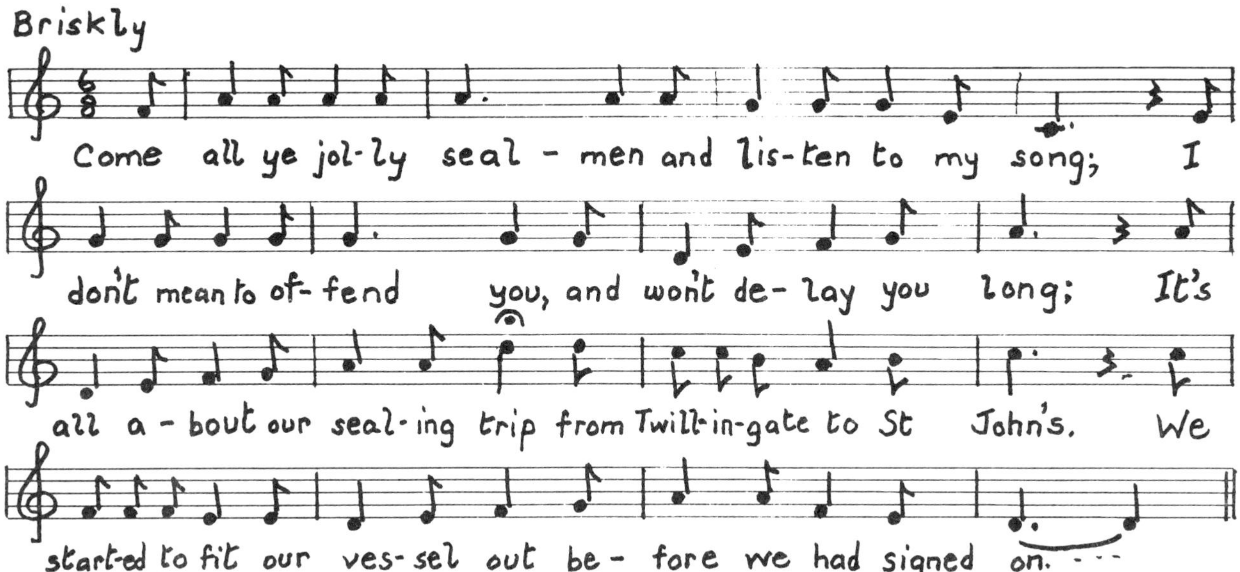

Come all ye jolly seal-men and listen
  to my song;
I don't mean to offend you, and won't
  delay you long;
It's all about our sealing trip from Twillingate
  to St. John's.
We started to fit our vessel out before
  we had signed on.

Our ship was fitted very well, from a radio
  to a shovel.
The only thing delayed our ship was a little
  engine trouble.
While taking in our ballast, some of us were
  drunk,
And more of us worked very hard, while the
  others lay in bunk.

It was on a Tuesday morning when our
  captain came from shop.
He said, "My boys, you'll now sign on, and
  then you'll get the crop."
Our crop composed of boots and clothes,
  likewise a fork and pan.
"If there's anything else you want, my boys,
  you must get it how you can."

Some of us took oilclothes and one of us
  took a watch,
He had it for to see the time, while he was
  at the swatch.
We enjoyed ourselves there very well, with
  laughter and with smile,
When Thomas White he went on deck,
  saying, "B'ys, here comes the ile!"

Our captain's name was Solomon White,
our chief mate was John Oake,
Our bo'sun was George Daley, a good man
for a joke.
Then tenth day of March, at dawn, from
St. John's we set sail,
With steam and canvas for the north she
covered her lee rail.

At four o'clock that evening we put her
in the ice;
We had to get her back again, and that
did not look nice.
On the following morning the captain
called all hands;
He thought it a good suggestion to put us
on the rams.

Northeast by east and east northeast her
course we steered that day,
Thinking to strike the whitecoats off
Bonavista Bay.
We motored in the daytime, and tied up in
the night,
And on the following evening the *Nascopee*
hove in sight.

The captain he did go on board and the
navigator too,
Reports fifty seals was on board, and all
well was her crew.
While listening to the radio, we received
good news that night;
The captain said he had to go, if the ice
was not too tight.

We motored until three o'clock, and then
we struck the fat,
Herbert Legge picked up a seal, Claude
Hawkins got a cat.
All hands went out upon the ice, to do the
best they can.
We picked up all our seals that day, but
minus of one pan.

We killed most everything we saw, from a
hood unto a harp;
I don't just know who killed the most, but
I think it was John Sharp.
Two accidents befell our crew upon that
very day,
When Robert Legge met a narrow escape
about two miles away.

Peter Trooke, a smart young man, was
working in the hold
When a cask of oil fell through the hatch
and gave him a severe blow.
Edmond Hines was a smart young man and
everything went well,
Until we donkeyed him five times and he
got mad as hell.

Our crew all numbered twenty-eight, with
seven in the watch;
Seven rifles were used among these men,
and they were all keen shots;
Now these two men I must include, and
that's the engineers, —
Herbert Watkins was our chief, Jack White
when he's not there.

It was on a Tuesday morning we made
another start,
When Gordon Dove cried from the barrel,
"I can see the schooner *Harp*!"
We steered our course for Bonavist', the
water calm and still,
But before dark we anchored in the place
called Wesleyville.

Seldom-come-by was our next port; it was
there we had to call;
The ice was cutting by the Cape, a knock-
back for us all;
We slipped our lines in Seldom, for north-
ward we were bound;
The ice was cutting by the Cape, and we
could not get round.

And one thing then we did spy out, that our
rudder was split in two,
It was Walter Pilky found it out, a benefit
to the crew.
And now to conclude and finish, I've one
thing more to say, —
It was about one mile from Seldom, where
we carried our blades away.

On the twenty-fifth of April, as we were
near our town,
Four rodney's we then put out to tow her
to the town.
Now our crew and captain must be men-
tioned, and I believe my song is the
longest of all,
And if you want a berth to the ice, please
give Mr. Ashbourne a call.

Reprinted with permission from Gerald S. Doyle, OLD TIME SONGS AND POETRY OF NEWFOUNDLAND. (St. John's: Gerald S. Doyle Ltd., 1940), 14.

An army of sealers returning to their ship. c. 1900. *Courtesy of the Provincial Archives of Newfoundland and Labrador (PANL).*

The Sinking of the "Newfoundland"

We have bred many seamen bold
Brave captains by the score
And ranking with the best of them
Is Captain John Blackmore.

He's one of nature's gentlemen
He does not rant or rave
He'll always greet you with a smile,
Though his thoughts be grave.

In early youth he went to fish
In the Straits and Labrador;
From Bonay Down to Ryan's Bay
The whole coast did explore.

As captain, he oft made two trips
During one fishing year;
He'd bring a load from out the Straits
And one from North somewhere.

Then when the fisheries did wane
He left his native town,
And at Port Union built a home
And there he settled down.

Not satisfied with sailing ships
He built the *Nanituck*
And sold her with idea in mind
To try again his luck.

The *J.H. Blackmore* then he built,
A sturdy craft was she;
He used her in the coasting trade
And in the seal fishery.

He lost her on a sealing trip
Near Elliston's rough shore;
He with his men they scaled the cliffs
And reached their homes once more.

Having then passed his three-score years —
His grit we all admire —
To build a sturdier, larger ship
Became his great desire.

Next winter found him with his son,
Working with heart and will,
Building the *Newfoundland* fair
In a dock at Clarenville.

He captained her in coasting trade,
Often in Arctic Sea,
And every spring he used her in
The North seal fishery.

Last year, he stepped down from the bridge,
His son, Fred, took command
Now, what this meant to Captain John
Sailors will understand.

To quit the sea and stay ashore,
It almost broke his heart,
And when the time for sealing came,
He had to take his part.

And so this spring, he sailed again
On his ship *Newfoundland*
Only to lose his ship once more
And have to walk to land.

For his ship was caught in rafting ice
On the morn of Patrick's Day,
With seven thousand young harp seals,
Crushed — on the bottom lay.

Now every man in Newfoundland
Does this misfortune rue,
And offers deep sympathy
To the Captain and his crew.

Yet when he reaches his home port,
Despite his loss, meanwhile,
He will remain undaunted, calm,
And greet you with a smile.

Reprinted with permission from Solomon Samson, A GLIMPSE OF NEWFOUNDLAND IN POETRY AND PICTURES. Edited by Robert Saunders. (Poole, England: J. Looker Ltd., printers 1963), 53.

The Barrack's Song

On Wednesday morning May the third
Nineteen and forty-four,
We left our homes seal hunting went
Rowed three miles from the shore
We met the ice twas very slack
Into it we did go
Expecting every moment to get a shot or two.

About two o'clock in the afternoon
We found that we were jammed
We knew it was no use to try
To get back to the land
So we poked into the water on the back of all the ice
And we rowed her for the barracks
And there we spent the night.

When daylight broke next morning
It was a dismal sight
To look towards our Island
And see three miles of ice
'Bout nine o'clock that morning
Two motors hove in view
With food on board to save our lives
Likewise two sturdy crew.

About four o'clock that evening
A little speck was seen,
To our surprise we soon found out
It was the *T-14*
She came right close in reach of us
The skipper gave a roar
Put your boat into the water boys
And I'll land ye safe on shore.

I'll tell you we made no delay
Oh we had no time to lose
And pretty soon we were on board
Relating all the news
We must give praise to the skipper
He is a sturdy man
Likewise, Lieutenant Oliver
Who tried to get a plane.

But we won't forget the people
Who prayed with all their might
Likewise the boys who lit the fires
And shot the guns that night
I hope twill never happen
To anyone again
And the author of this little poem
His name is Nicholas Lane.

This song was recorded in 1973 by John P. Green from Mr. Nicholas Lane of Tilting, Fogo Island. Mr. Lane, who is in his eighties, sent us a copy of the song in his own handwriting. It included only a few changes from the tape recorded version. Printed by permission of John Greene and Nicholas Lane.

Hunting Seals

With knife and fork, with kettle and pan,
With spoon and mug, and glasses
To shield our eyes from glaring sun,
And we take our tea with "lasses."

Cho. — For we are swoilers fearless, bold,
As we copy from pan to pan, sir;
With pelts astern we shipward go —
Nor yield to any man, sir.

With sheath and steel strapped to our waist,
With a stocking filled with bread, sir,
We leave the side away to stride
O'er the icefields without dread, sir.

Cho. — For we are swoilers fearless, etc.

With bat and gaff, and "panning staff"
Surmounted with a flag, sir;
Away we go on the great iceflow,
And we never care to lag, sir.

Cho. — For we are swoilers fearless, etc.

With a tow line round our shoulders bound,
Steel "frosters" in our boots, sir,
To aid us in our hauling, to
The pans that hold the lot, sir.

Cho. — For we are swoilers fearless, etc.

With harps to play on Patrick's day,
We all take great delight in;
But with the hood there's no such fun,
For they're the devils for fighting.

Cho. — For we are swoilers fearless, etc.

Sometimes a Polar Bear we'll meet,
But of them we are as charry,
As of the maid we left behind,
Who'll this year ask to marry.

Cho. — For we are swoilers fearless, etc.

Of bedlamers we often take,
Their "pelts" being quite as good, sir,
As any "swoil" in yield of oil.
Be he "dog harp", or hood, sir.

Cho. — For we are swoilers fearless, etc.

Again, a white coat we will get,
To make the girls a muff, sir;
If she turns up her pretty nose —
The taxidermist — he'll stuff, sir.

Cho. — For we are swoilers fearless, etc.

And when we've loaded hold and deck
With pelts and plenty flippers,
We leave the floe and homeward go
To meet delighted skippers.

Cho. — For we are swoilers fearless, etc.

Reprinted with permission from an unidentified manuscript, believed to be the work of James Murphy, and deposited in MUNFLA by Dr. John Widdowson of the University of Sheffield. Courtesy of Fred Earle, Memorial University Extension Service.

Air: The Rigs O Barley.

Note. — The following song shows the methods displayed by the hardy Newfoundland sealer when hunting for the treasures of the ice floes.
(Original note)

The Fisherman's Son to the Ice is Gone

The fisherman's son to the ice is gone,
On the quarter deck you'll find him;
His belt and sheath he has girded on,
And his tow rope slung behind him.

The ice slacked off we got our load,
Our good ship homeward steering;
Flushed with success our bosom glowed,
As St. John's we are nearing.

Blow gentle breezes soft and clear,
O'er the bright blue waters;
And bring us to those that we love so dear.
Fair Terra Nova's daughters.

Reprinted with permission from James Murphy, SONGS SUNG BY OLD TIME SEALERS OF MANY YEARS AGO. (St. John's: James Murphy, 1925), 7.

The Girl I Left Behind Me

"I'll go to the ice and I'll catch some seals,
And all the crew will join me;
And when I come home, I'll bring a bill
For the girl I left behind me."

Reprinted with permission from James Murphy, SONGS SUNG BY OLD TIME SEALERS OF MANY YEARS AGO. (St. John's: James Murphy, 19.25), 19

The Sealing Fleet

What means this hurrying to and fro —
This busy stirring scene?
Is it a great and gala day
For England's noble Queen?
Or is it a proud spectacle,
Where deeds are to be done,
That shall excite the mother
For the glory of her son?

The scene laid now before you
Is not of war nor strife:
But 'tis a fight of honest men
To nourish mortal life,
They go to catch the northern seal,
To buffet sea and storm:
May God protect those noble men,
And keep them from all harm.

Gaze on those Terra-Novan sons,
Adventurous and bold:
There's nobleness in every look,
The youthful and the old;
Some are husbands, and they go
With wives' pure, fervent prayer:
The poor young children have to wait
For their dear father's share.

See that youth of ruddy face,
With bold and manly tread:
He has no mother to console
His father too is dead.
His fate has been a lonely one,
Though his looks are quite resigned;
Both parents died when he was young —
He is an orphan child.

Poor youth, the world forsaketh him,
A stranger he is now;
But he may yet win laurels,
To deck his humble brow.
If truth and honor be his guide,
He'll scorn the cold world's sneers;
He works now for a noble aim, —
She'll soothe his future years.

And we, the quiet lookers on,
Who stop at home at ease,
Are watching now with beating hearts
Your sails furl in the breeze.
May God fill all your little crafts
With weighty polar seals,
Is the wish of every honest man
Who for his country feels.

Garryowen

Reprinted from THE NEWFOUNDLAND EXPRESS, March 5, 1861. Courtesy of Dr. Cyril Byrne.

Landsman returning home with his "tow". *Photo by Fred Earle, Memorial University Extension Service.*

Arctic Ice and Flippers

There's a halo round the margin of the sea,
And 'tis there, if I correctly guess, will be
The Arctic Ice — the whelping ice whereon the "fat" is found;
The haze is thick and stretching 'cross the Bay;
And scattered pans of drifting ice are floating all around.
We'll get the flippers yet old-timers say.

There's a strong Nordeaster blowing cold and raw.
More birds along the shore I never saw.
The ducks and hounds are flying, and the turrs are near the land;
You can see the odd young "beater" in the Bay,
The *Terra Nova's* off the Cape with seven thousand panned.
We'll get the flippers yet old-timers say.

How far the floe is off it's hard to tell:
The surest way to judge is by the swell;
'Tis running high, and roaring loud upon the Nordeast shore;
No sign of moderating yet today;
I'd guess the ice is off the land 'bout twenty miles or more.
'Twill fill the Bay to-night old-timers say.

So get your gaff and tow-line ready, John,
And meet me on the Point at crack o' dawn!
We'll be first upon the "growlers" from the Shag Rocks to Cape Freels
And strike the "fat" before another sinner.
You can bet your bottom dollar we will get our tow of seals;
And it's flippers fat and sweet for Sunday's dinner!

This song was written by Mr. A.C. Wornell in 1937. It is reprinted here with his permission from A.C. Wornell, RHYMES OF A NEWFOUNDLANDER. (St. John's: Guardian Limited, 1951), 15.

The Old Time Sealers

The bells they are ringing, the sirens are screaming,
The flags they are flying from ship and from shore,
  The wind it is blowing,
  The snow it is snowing,
The sealing fleet's ready to leave port once more.

Just look at them clouds, there is plenty of snow there,
A wicked Nordeaster is raging outside,
  We'll anchor midstream,
  Through the Narrows we'll steam
When the ice goes to seaward upon the ebb tide.

Make ready your ropes and your gaffs and your oilskins,
The man in the barrel is keeping his watch,
  We'll wait for the word
  That he's sighted the herd,
Then it's over the side and we'll strike the main patch.

The ice is alive with the young baby whitecoats
They're wailing like infants as over we go,
  A knock with the bat
  Then you rip off the fat,
In a little while now I'll be scoting me tow.

This life is a hard one, I'm sure you'll agree
But we really don't mind when we're sure of our pay,
  There's pork with the duff
  And if that's not enough
Sure they give us molasses to sweeten our tay.

A man snogging seals must be sturdy and strong
And you certainly work for the money you earn
  There's jumping and running
  There's knifing and gunning
There's danger awaiting wherever you turn.

You're up with the dawn and you work half the night
First you pick up the pans then you stow them below,
You trudge many miles
When you're hunting the swiles
And you often drag three hundred pounds to a tow.

We're headed for home and we're loaded with fat
There's a dozen of flippers put by on a string,
We're tired of hard tack
But we'll gladly go back
For a berth to the ice with the sealers next spring.

Don't tell me about men who go mountain climbing,
Don't mention the dangers of hunting big game,
We'll sing and we'll shout
For there isn't a doubt
That the Newfoundland sealer could put them to shame.

Reprinted with permission from NEWFOUNDLAND STORIES AND BALLADS 10:2 (1964), 33.

The Old-Time Sealing Fleet

Newfoundland has many stories that can make the heart beat fast,
Stirring tales of great adventure from the annals of the past;
But to grip and hold the fancy, there is nothing to compete,
With the long and thrilling sagas of the old-time sealing fleet.

How it fired my boyish spirit, how it made my heart rejoice,
As some gray old-timer told me of his last spring to the ice;
And my thoughts would bear me onward with the captain and the crew,
To the icy wildernesses, and my pulses beat anew.

Where the blinding snow-storm threatens, rushing on in fiendish glee,
And the moving isles of winter, shock with noises of the sea;
Where the white-coat whimpers sadly for the murdered mother-seal,
And the vessel staggers blindly as the "growler" strikes her keel!

Well the sealer's heart may tremble, when he sees the falling "glass",
And the giant bergs are tossing fragments from their mighty mass;
Howling night comes on, and crashing round the trapped and stricken ship,
Ghostly monsters heave and topple with destruction in their grip.

But when first the seals are sighted, cradled by the icy pan,
Former dangers are forgotten, and the heart of every man
Leaps with sudden exaltation, as he gazes on the sight,
And he deems it ample payment for the terrors of the night.

Then their loud exulting voices rise toward the northern skies,
And the shivering seal's low moaning mingles with the hunter's cries;
Now the youthful sealer listens, and his eye with pleasure gleams
For before him is the haunting "El Dorado" of his dreams.

There are countless tales of daring to be heard with bated breath,
Of the plunge in icy waters and the comrade snatched from death;
And a thrill of admiration stirs within us as we hear,
Mighty feats of strength recounted from the days of yesteryear.

There are stories too, of shipwreck that a sealer can disclose,
Of the tragic death of comrades as they battled with the floes;
For the stern accountant Nature, heeding not the stricken wives,
Sometimes balances her losses with the toll of human lives.

As we listen to the sagas we are proud that we can say,
That the land which bred those heroes can produce the same today;
Still, as winter snows forsake us and the spring comes round again,
Iron ships, go forth to battle, manned with crews of iron men!

Still the Arctic tide flows onward, bearing on its swelling breast,
Mighty bergs and moving ice-fields, where the precious whitecoats rest;
Still the tempest howls in fury with its storms of blinding snows,
And the old dog Hood is waiting for the coming of his foes!

Newfoundland has many stories which can make the heart beat fast,
Stirring tales of great adventure from the annals of the past;
But to me as I review them there is nothing to compete,
With the long and thrilling sagas of the old-time sealing fleet.

Reprinted with permission from A.R. Scammell, SONGS OF A NEWFOUNDLANDER. (St. Jerome, P.Q.: J.H. Labelle, 1945), 11.

Last of the Wooden Walls

Here Atlantic's foam-wreaths float
In aqua-floral tribute to a ship submerged;
A wooden ship whose more than forty years of service
Played no meagre part designing and developing
The economic pattern for an isle
Versed in maritime activities;
A ship whose Captains sealed their fame in sealer-strategy,
And often risked the sealing of their fate;
Whose Barrelmen, aloft where wind-bowed riggings
played
Incessant melody,
Heralded the whitecoat patches, bedlamers and old dog-
hoods
Out on the frozen front —
Out on the Arctic Floe;
Whose hardy officers and crews,
In order not to lose their springs,
Trudged through miles of snow-drifts
Along rugged coastline interspersed with rugged
outports,
Whence they came;
Withstood thc northern elements of many a whitecoat
season
To "swoile" the crusted swell.

And, oh, what frigid terrors
Seized those hunted creatures,
Harmless and comparatively unarmed,
While ruthless each battle
Gaffed its ghastly toll,
The spoils of which pelt-bulged the vessel's holds;
How piercingly pathetic were those whitecoat whimpers,
Those vivid, mercy-pleading eyes
That would melt the heart of ice —
Pleas, alas, essentially denied.
What valorous events occurred
Within the bleak perimeter
Of this vessel's often grim and lone location!

Events now logged in long-spun yarns
And the annals of legend and sealer-song
And sometimes there was time for sealer-song
In the inimitable rendition
Peculiar with her complement,
Accompanied by strains of rafting seas.

And during storms,
Only the breath of prayer had animated hope,
Consoled the anxious ones at home —
Those whose hearts had beaten gale-accelerant time,
Whose eyes had peered eastward
Through rough-patterned panes;
But, majestic like her name,
This vessel bore and breasted storms,
Brought her gains to port,
That citizens could feast on flipper pie,
And crews receive their wages
And trudge back home —
Back where high hopes had been stored
In that hard-earned bill:
Little Jack, who had yearned to copy pans all spring,
Would have had "them dandy boots"
And Jane would not have had to work
Too hard and long for Mrs. Smith;
In fact, they would have been sent to school next term
But then there had always been the doctor's fee —
The reverend to pay...

But here Atlantic's foam wreaths float,
Here a kelp-enshrouded heirloom
Unravels in aquatic coffers;
Through the throes of pomp and pathos,
She had slowly crossed the bar,
Bowed farewell to her harbour port
That reluctantly had bidden sad adieu,
Bowed down her hundred fathoms...

And in her sinking spell,
More tears had flowed than man cares to tell.

By Harry R. Burton

Reprinted with permission from Michael Harrington (ed), POEMS OF NEWFOUNDLAND. (St. John's: F.M. O'Leary Ltd., 1952) 90-92.

The Swiler's Song

Chorus: Rise up me hearties with gaff and sculp
With hobnails rope and line
To hunt the swiles all in the gulf
On the ice in the cold spring time
We'll live on tea and stale hard tack
And work till our backs are sore
But a bumper load and the wind at our back
Will bring us home once more

Step lively now me bully boys we've got some miles to go
The journey won't be easy but no one told ye so
No no one ever told ye the journey would be nice
But the walking that ye do on land will help you on the ice

From every bay and headland and all points in between
It is the hope of every man to ship with Old Man Kean
Which ever ship ye draw boys I'm sure she'll suit ye well
For one is like the other and the fleet is a floating hell

Take off your caps me bully boys and wave them in the wind
To cheer our wives and sweethearts that we do leave behind
And now that we are berthed and settled for to sail
'Tis a hope 'twill be the main patch that calls us o'er the rail

Lift up your heads me hearties and do not be downcast
For getting to the swiles boys is a thing that's not done fast
So dig your hobnails in boys and lend the rope your backs
And scamper like the devil the minute that she cracks

That roar all from the topmast it is a welcome sound
We'er in the swiles me hearties and there's whitecoats all around
Swing o'er the side right lively and clamper o'er the pans
There's pelts to line your pockets and blood to warm your hands

So sculp 'em quick me buckoos and pile them by the flags
Then go to find another patch over pinnacles and crags
There is no time to tarry the day is getting late
Take what ye can today boys for tomorrow they'll not wait

Turn round turn round me hearties we have a day's work done
There's weather in the northeast and dogs around the sun
Let's all join in a song boys to keep our spirits high
But listen for the whistle and keep your stockings dry

Trust in God almighty and in our captain's ken
For you know he won't abandon hardworking swilin' men
But if our time is come boys there's nothing we can do
For many a hearty swiler sleeps 'round the Funks and Baccalieu

This song was composed by Pat Byrne and is printed here with his permission. Melody, "The Sinking of the Caribou".

Captain Abram Kean

We should not forget the Commodore,
The old king of the sealing fleet.
For grinding floe and tempest's roar
Were part of his bread and meat.

With unerring aim and judgement rare
He would strike each sealing patch.
'Till at length we heard the records declare,
There's a million down the hatch.

For fifty years he butted the ice,
And went through his woes and weals.
He was often gripped in an Arctic vise,
But he landed a million seals.

And a million seals is the highest score,
It will never be done again;
So we should not forget the Commodore,
The late Captain Abram Kean.

Reprinted with permission from Otto Kelland, ANCHOR WATCH: NEWFOUNDLAND STORIES IN VERSE. (St. John's: Dick's & Co., 1960), 88.

Ballad of Sealing Ships and Sealers

Come all ye hearty Newfoundlanders, join your voices now with me:
Of our sealing ships and sealers let us sing:
Of the men who hunt the white-coats on the ice-encrusted sea
And endure the blinding blizzards every Spring.

With their whistles piping loudly in response to shore farewells,
With their code of signals flying fore and aft,
The flotilla, steaming slowly, heads toward the ocean swells,
As they leave historic Signal Hill abaft.

Soon their sturdy bows will shiver as they butt the Arctic floe,
And their sides, encased in greenheart, then will groan;
Whilst the agitated "growlers", rafted by the undertow,
Make a melancholy music all their own.

Persevering in their venture, they will zig-zag to and fro,
As the "look-out" in the barrel "scuns" them through
To the "leads" of open water, steaming ever Northward Ho,
Till the whelping-ice is welcomed into view.

Then will sound a joyous chorus from a hundred sealers throats:
"Seals ahoy! We've struck the patch!" they'll loudly sing:
Whilst the skipper ponders proudly, as upon the scene he gloats,
That they'll get a bumper trip for sure this Spring.

With their tow-lines on their shoulders, and their knives stuck in
their belts,
With their grub-bags on their backs and gaffs in hand,
The excited crew await the word to go and get the pelts,
And they'll soon have several thousand flagged and panned.

But their work is not all pleasure, for the weather's often rough;
And the ice can crush the stoutest ship that floats;
It's an occupation calling for a constitution tough,
Not a job that can be done by "petticoats."

So let's wish them all good sailing, from the mess-boy to the mate,
With a special prayer for captains in command:
For without their skill and daring, there'd be little to relate
Of the men who hunt the seals in Newfoundland.

Arts and Letters Contest 2nd Prize — 1954
(Original note)

Reprinted with permission from A.C. Wornell, RHYMES OF A NEWFOUNDLANDER. (St. John's: Guardian Limited, 1951), 9.

The Dying Seal-Hunter

I can hear their sirens blowing
As they steam to hunt the foe,
Where the young whitecoats are growing,
As they bed down in the snow.

Raise my up that I may see her,
My old ship, and a last farewell
To her captain by the wheelhouse,
As she clears the harbour bell.

How I long to be out with them,
Faithful comrades every one,
Where the diadems of Arctic
Glisten in the morning sun.

Where the North King reigns in splendour,
As they beat into his realm,
Sturdy Newfoundland seal-hunters,
Masters of ice-pan and helm.

As they fade out from my vision
Down the Northern Front to roam,
Fate has made for me decision,
I won't greet them coming home.

Raise me up that I may see her,
My old ship, and a last farewell
To her captain by the wheelhouse,
As she clears the harbour bell.

Reprinted with permission from Otto Kelland, ANCHOR WATCH: NEWFOUNDLAND STORIES IN VERSE. (St. John's: Dicks & Co., 1960), 90.

The Sealers of Twillingate and New World Island

True are the tales which fishermen tell,
And stories they relate.
Of dangers faced and overcome,
In outport Twillingate.

For many years you've gone upon
The ice-floes far and near.
And history records the feats
Performed throughout the years.

Three hundred years have come and gone,
Since men did immigrate
From England's soil, across the waves,
To live in Twillingate.

They've gone out on the mighty deep
When storm clouds hid the sky,
And wrestling with the elements,
Some brave men had to die.

An island home, close by the sea
Atlantic Ocean blue,
Has brought much wealth upon the floes,
Since eighteen sixty-two.

That was the year of the "Big Haul"
Men walked out from the shore.
They toiled for weeks and brought to land,
Ten thousand seals and more.

The seals "struck in" St. Patrick's Day.
When fiercely blew the wind,
And eager men with gaff in hand,
The harvest did begin.

They toiled all day from early dawn,
'Til darkness filled the sky.
Two sealers from exhaustion fell,
Upon the ice to die.

Two more were driven far from land,
While on the ice they roamed,
It was a tragic price to pay,
With loved ones left at home.

One widowed damsel, plucky girl,
Her fam-i-ly to feed,
Went out herself to hunt for seals,
To help supply their needs.

A special mention she deserves
(She bravely did her thing)
By hard work, pluck, she earned that year,
One hundred pounds sterling.

The years have passed since '62
But men still have the will,
When winds and tides are favourable,
To hunt the young harp seal.

With frost severe and winter storms,
The season did begin,
And veteran sealers did predict,
The whitecoats would come in.

And now 'tis nineteen seventy three,
Although the times have changed,
With bats and gaffs and powered boats,
Far from the land the range.

From Too Good Arm and Summerford,
From Moreton's Harbour too,
The sealers went to Sleepy Cove
By car, truck and skidoo

In fact they came from all around,
Excitement running high,
The older men had never seen,
Such scenes when they were boys.

March-April days were busy ones,
Men took off from the shore,
And brought to land by rope and boat,
Eight thousand seals and more.

While Brian Davies buzzed o'erhead,
(He's from S.P.C.A.)
He chose to fly aloft to snoop,
Or he'd been clubbed one day.

This foolish lad from upalong,
Has notions in his head,
He'll stop the kill of young harp seals,
And let them live instead.

He's asking folks to sign their names,
To send to Ottawa,
But Parliament, with Lundrigan,
Will not pass such a law.

The harvesting of young harp seals
Is for the people's good,
Besides the cash which circulates,
The flippers make good food.

What were the prices paid for seals?
What did the sealers get?
Two men for one large old harp seal,
Thirty-five dollars net.

Ten bucks were paid for beater seals,
With white-coats fetchin' seven,
They were not bought by weight or size,
For round seals this was given.

Yes, history does repeat itself,
The future will relate,
This was the spring of '62,
Again in TWILLINGATE.

John C. Loveridge

The Seal Hunt of 1862 and 1973.

Reprinted with permission from John C. Loveridge, STORY IN PICTURES AND POETRY OF THE 1973 SEAL HAUL AT TWILLINGATE NEWFOUNDLAND also MARKING THE 85th YEAR IN BUSINESS OF THE FIRM OF E.J. LINFIELD. (Grand Falls: Robinson Blackmore, 1973), 3.

A whitecoat peers out from behind a knob of ice. *Photo by Hal Andrews.*

A Sealer's Reply to His Wife

Now that March month has come
And spring's in the air,
The old seals are swimming
Up North to their lair.
And soon on the ice-fields
Their young will be strown,
So Maggie my darling
I must leave you alone.

For my berth to the ice
I cannot turn down.
I must join the sealers
And go into town;
To sign on a sealer
To the ice-fields that's goin'
So Maggie my darling
I must leave you alone.

The call of the ice-fields
Comes forceful and clear
It makes me feel restless
At this time of year.
I long to be heading
Where north winds have blown
So Maggie my darling
I must leave you alone.

Now the price of fat's down
And so is the fur,
But from this adventure
It can't me deter
In the thick of the ice-flow
We'll soon now be thrown
So Maggie my darling
I must leave you alone.

There's fur and there's danger
A hunting the seals.
For you jump from the ship
As she shivers and reels.
Over rough moving ice
You run or lie prone,
But Maggie my darling
I must leave you alone.

And when we return
With a good bumper trip,
You will soon forget
That I gave you the slip.
And when I'm too old
To the ice to be goin'
Then Maggie my darling
I'll not leave you alone.

A sealer's reply to his wife who tries to dissuade her 60-year old husband from going to the icefields.
(Original note)

Reprinted with permission from Solomon Samson, A GLIMPSE OF NEWFOUNDLAND IN POETRY AND PICTURES. Edited by Robert Saunders. (Poole, England: J. Looker Ltd., printers, 1963), 41.

Uncle John — The Sealer, 1951

Among the sealers who came home
From the seal fishery
Was stocky wind-tanned Uncle John,
As mad as he could be.

For over forty years and more
A sealer he has been.
He says "That killing infant seals,
He never 'fore has seen.

'Tis naught but bloody murder boy,
Of which I've had my fill
And if Bradley don't try stop it
I hope Gordon Higgins will.

Why have our sealing laws been sheared
Is what I want to know.
For March the thirteenth was the date
You first could kill a tow.

But now as soon as they are pupped,
Before they're any good,
You're forced to murder in cold blood
The baby harp and hood.

If Mr. St. Laurent could see
What I have seen this spring
'Twould give him nightmares in his sleep.
He'd stop the horrid thing.

For until our old sealing laws
Are put in force again
Every true native mother's son
Should bow his head in shame.

And Uncle John will never rest,
Nor will his temper wane,
Until protection for babe seals
Is made the law again."

(In 1952 a law was enacted making March 10th, the earliest date that young seals could be taken).

Reprinted with permission from Solomon Samson, A GLIMPSE OF NEWFOUNDLAND IN POETRY AND PICTURES. Edited by Robert Saunders. (Poole, England: J. Looker Ltd., printers, 1963), 39.

We Will Always Have Our Sealers

We will always have our sealers
While there's a ship to sail,
While sturdy crews have fish and brewis,
While there is rain and hail.
We will always have our sealers
While there is pork and duff,
While flippers rove the galley stove,
Where green seas cut up rough.

We will always have our sealers
While danger haunts the floe,
While whitecoats crawl and old harps bawl,
Our lads will ever go.
We will always have our sealers
While there's an old dog hood,
While sharp prows ram the stubborn jam,
To the tune of tortured wood.

We will always have our sealers
While there's a gaff or flag,
While men get welts from haulin' pelts,
While there's a tow to drag.
We will always have our sealers
While there is snow and ice,
While frost-bite pales from Northern gales,
Where men and death throw dice.

We will always have our sealers,
New industries may thrive, —
Though Arctic mail sheathes the win'ard rail,
And growlers crush and drive.
We will always have our sealers
Though ice packs raft and shove,
Then when life's through they'll start out new,
And sign on up above.

Reprinted with permission from Otto Kelland, ANCHOR WATCH: NEWFOUNDLAND STORIES IN VERSE. (St. John's: Dicks & Co., 1960), 89.

Save Our Swilers

Come all you Newfoundlanders and listen to my song
About St. Anthony's visitors from "away" and "upalong";
There were movie types and media types and Mounties, some fivescore,
If we were bent on violence they'd need a hundred more.

They are out to ban the seal hunt and this they mean to do,
Brian Davis and the Greenpeacers and all their motley crew;
This year they've got Franz Weber with phony fur to sell —
A bleeding heart from Switzerland who thinks he's William Tell.

They say the seals are threatened but the evidence is clear,
With quotas carefully controlled, of that there is no fear;
We're the endangered species who live by coastal seas,
We kill the seal as we kill fish to feed our families.

A bedlamer boy from Greenpeace he chained on to the "whip",
And was dunked into the water by the rolling of the ship;
We had a job to save him in all the fuss and racket,
But I bet his pelt wouldn't have been worth as much as a Ragged Jacket.

When Brigitte said in Paris she cuddled a whitecoat dear,
Sure every swiler in the land he grinned from ear to ear;
He knows from long experience she's pilin' on the lies,
A real whitecoat's talons would have slashed her face and eyes.

They call us cruel, barbaric, hunting seals just for the thrill
These pampered city slickers that a day's hard work would kill;
What do they know of challenges of storm and sea and ice
That dare the blood to answer and to pay the sealers' price?

They're out for front-page stories, they've come so far to roam,
And blood on ice will show up well on T.V. screens back home;
They know their media bosses have paid good money out,
If they don't send "juicy" stories their jobs are "up the spout".

There's many things we don't approve in countries far away
How people act and dress and talk and how they earn their pay;
But we don't get up a hate campaign and stir up children too,
To force our views on other folks as these do-gooders do.

Our government must keep these types from off the whelping ice,
Or there'll be tragedies to tell — we're men, not frightened mice;
We merit more protection than a motion on the floor
Where will you get your slippers when the seal hunt is no more?

We're not averse to meeting up with a star from Hollywood,
We sure would like to rescue her from an angry old Dog Hood;
But the Arctic floes are not the place, Yvette, to use your wiles,
We're not spruced up for courtin' when we're out there peltin' swiles!

We have to take from Nature whate'er the season bring,
We're fishermen in summer and swilers in the spring;
If you don't approve the seal hunt, you have a right to say,
But when we go out on that ice, don try to bar our way.

So here's a health to Romeo, who took the sealer's part,
He stood up to protesters, he has our cause at heart;
And raise your glass to Tommy Hughes who tells it like he knows,
And don't forget Rick Cashin when you're culling friends from foes.

By A.R. Scammell

A Sealers' Song — 1977.

(Tune "Old Polina)

Reprinted with permission from DECKS AWASH 6:4 (June, 1977), 6.

NOTES ON THE SONGS

For the most part, the notes given here are references to versions of the songs other than those found in the text. We include in each citation the authors name, date of publication, and page reference. The complete citation may be found by consulting the bibliography.

Abbreviations found in the notes are:

NSB: NEWFOUNDLAND STORIES AND BALLADS

MUNFLA: MEMORIAL UNIVERSITY OF NEWFOUNDLAND FOLKLORE AND LANGUAGE ARCHIVE

WADHAMS SONG
Byrnes 1931:147; Doyle 1927:35; Doyle 1966:70; Greenleaf 1968: 236; Murphy 1905:22; Murphy 1923:14; NSB 4:1 (1957), 9; Smallwood 1937:369. This is not a sealing song but was performed as a recitation giving direction at sea, heading north from Bonavista to Notre Dame Bay.

THE SEALER'S CALL
Samson 1962:14..

CAPTAIN HENRY THOMEY
L.G. Chafe writes about Captain Henry Thomey:
He was selected as Master for one of the sailing vessels before he was twenty years of age, and continued until he was nearly eighty years of age, and rarely failed to bring in a good trip. For an account of the rivalry between Thomey and Asariah Munden of Brigus see an essay by H.F.S. Shortis in the *Evening Telegram*, March 28, 1913. Follow up letters by men who served with them are found in the *Evening Telegram*, April 10 and 16, 1913.

WE WILL NOT GO TO WHITE BAY WITH CASEY ANY MORE
A variant is found in MUNFLA (68-23).

COME ALL YE JOLLY ICE—HUNTERS
NSB Vol. 14:2 (1968), 21; Greenleaf, 1968:244. We have been informed by Mr. John P. Greene that this song was written by Mr. William Greene of Tilting, Fogo Island. He wrote many songs and was known locally as a great singer. He was born on July 23, 1820, and died May 17, 1898.

OLD TIME SEALERS' SONG
Murphy 1902:74; Murphy 1905:20; Murphy 1923:1.

SONG OF *THE JOHN MARTIN (1)*
Doyle 1927:25; Smallwood 1937:471.

THE JOHN MARTIN (2)
See footnote citation.

FERRYLAND SEALER
Fowke 1967:62; Fowke 1973:50.

THE LAND OF FISH AND SEALS
Byrnes 1931:143; Murphy 1902:79; Murphy 1905:19.

DRIVEN INTO SPANIARD'S BAY
See footnote citation.

COME DOWN WITH THE KILLOCK
See footnote citation.

TERRA NOVA SEAL FISHING
See footnote citation.

A NEWFOUNDLAND HERO
Greenleaf 1968:294. Greenleaf comments on the song: "They got safe ashore and the women recovered, but his (Jackmans) health was broken. Queen Victoria sent him a medal. The date of the wreck and rescue was October 29, 1866." The song is also found in Smallwood 1967:466. C.W. Andrews and G.M. Story comment about Jackman: Born 20 May 1837 at Renews, Newfoundland, eldest child of Captain Thomas Jackman, a descendant of a native of the Isle of Wight who was shipwrecked on the southern shore of the Avalon Peninsula...entered the cod and seal fisheries as a boy with his father...between 1867 and 1876 he was in charge of Bowring Brothers' sealing steamers *Hawk* and *Eagle*...His death at the age of 39 cut short a sealing career of great promise....(Andrews and Story 1972:378).

CAPTAIN BILL RYAN LEFT TERRY BEHIND
A fragment of another version of this song is given in In a personal communication Dr. Story states that the Captain White (who was, by the way, Story's great grandfather) of the fragmentary version took the *Nimrod* to the ice for Job's in 1867, 1869 and 1870.

SEALERS' SONG
Bennett 1972:9; Blondahl 1964:73; Doyle 1966:44; Mills 1968:40. For different versions see Eckstorm and Smyth 1927:324; Peddel 1904:30; Peddel 1908:47; MUNFLA C883; MUNFLA C32. Captain Blandford referred to in the song, was a man who worked his way up from a very humble station, but by his pluck and indomitable skill became a leader of men, both at the seal and cod fisheries. He was also a member of the House of Assembly (Chafe 1923:31). Captain Barbour, also mentioned in the song, was in charge of the *Nascopie*, one of the largest and most powerful iron-clads on her first trip to the ice fields (Chafe 1923:32).

THE SEALING TRIP OF THE *S.S. GREENLAND* 1891
Murphy 1925:8. See Chafe 1923:46 for the same version. The captain mentioned in the song is Capt. Henry Dawe. He was not alone the most popular Captain at the front, but success followed him everywhere. If he missed the young seals he came in with a load of old ones. He was selected as Master of the *Adventure*, the first of the iron-clad, and made a wonderful success in that ship (Chafe 1923:31).

TRINITY BAY TRAGEDY
Leach gives the following note on the 'Trinity Bay Tragedy.'
This ballad describes a sea disaster that took place in Newfoundland in 1892. By way of contrast and explanation I append an account of the tragedy as written by a historian, D.W. Prowse, and printed in his HISTORY OF NEWFOUNDLAND (London, 1896), p.520. (original note)
"Saturday, the 28th of February 1892, is a dark day in the annals of Trinity, a day to be remembered and mourned. The morning sky ushered in a lovely dawn, the sky was clear, a soft, bright, balmy air blew from the land over the treacherous sea, the light breeze scarcely ruffled its bosom. From Trinity and every harbour adjacent, boats were out by early dawn in pursuit of seals, which had been seen the previous evening. From Trinity, Ship Cove, Trouty, English Harbour, Salmon Cove, and other small places, the daring ice hunters set off with high hopes and buoyant spirits to chase the wary seal; in this most exciting and dangerous pursuit the Newfoundlander recks not of danger; difficulties and perils that would affright one unaccustomed to the ice fields are mere sport to the hardy native. On this eventful day seals were few and scattered; in the fierce excitement of the chase many went far out in the bay, heedless of the coming tempest; a few of the older fishermen, especially those from Trinity, more wary, and probably less vigorous, noticed the first signs of the storm, and before the icy blast came down with full force they were under the lee of the land and could row in. Two hundred and fifteen men were out that day; the majority got safe to land after a tremendous struggle for their lives; the rest of the heroic fishermen in spite of their exertions were finally overpowered; with strong arms they rowed for their lives, but the freezing icy tornado swept down upon them and paralyzed their efforts; they had done all that men could do against the blizzard; they fought with the gale while instant death appeared on every wave. One bold crew from English Harbour seeing all their attempts to stem the tempest were in vain, made for the ice; so chilling was the blast that one young fellow became paralyzed with the cold before they reached the ice. However Newfoundlanders in a difficulty are never without resource. They climbed onto a pan higher than the rest, where they made a rude shelter; their boat was broken up to make a fire; with this and some seals they managed to live through that awful night. Thirteen fishermen were found frozen to death in their little punts; eleven others were driven up the bay and perished in that awful dark cold night of death...a gallant crew from Heart's Delight saved the sixteen men on the ice; the generous Captain Fowlow put out in the schooner and succoured some and brought them home to their agonized families. Charity flowed in to the widows and orphans...sympathy was called forth for the mourners, for those, alas, whose sorrow for the dead will never die, who all life long will grieve over the death and destruction of that dark day of storm."

DIED ON THE ICE FIELDS
Murphy 1905:13.

LOSS OF THE *MAGGIE*
Murphy 1905:5.

THE WRECK OF *THE MAGGIE*
See footnote citation.

THE SPRING OF '97
Peacock 1965:976.

THE *GREENLAND* DISASTER (1)
Peddel 1904:18; Peddel 1908:35; MUNFLA C522. M.A. Devine wrote a song about the *Greenland* Disaster entitled "The Dead Seal Hunters" which is included in McGrath 1974:13. Mr. Devine's version is also included in Smallwood 1937:358.

The *Greenland* Disaster (2)
MUNFLA C86; MUNFLA C690; Greenleaf 1968:299. For the same version of this song collected as a recitation see MUNFLA C298.

THE *GREENLAND* DISASTER (3)
MUNFLA C 298; Mowat 1973:92. This same version was collected by MacEdward Leach in 1950 from Jim Rice of Cape Broyle. (MUNFLA Tape 7A, Note Book 1, P.7. — Courtesy of Ms. Julie Hauri and Dr. Kenneth Goldstein).

WRITTEN IN MEMORY OF THE 48 MEN LOST IN *GREENLAND* SEALING DISASTER MARCH 21, 1898 (4)
See footnote citation.

GREENLAND DISASTER (5)
See footnote citation.

THE *GREENLAND* DISASTER (6)
See footnote citation.

THE *GREENLAND* DISASTER (7)
Wornell 1972:31.

THE ICE FLOES
D.G. Pitt says that this is an imaginary disaster at the ice-fields, but is based on fact. In 1898 forty-eight men from the sealing-ship *Greenland*, and in 1914 fifty men from the *Newfoundland*, perished in circumstances similar to those described in the poem (Pratt 1962:161). We have been also informed by Dr. David Pitt that E.J. Pratt's grandfather was a sealing captain and once lost a vessel off Cape Bonavista.

SEALING FIFTY YEARS AGO
See footnote citation.

THE SEALER'S STRIKE OF 1902
Murphy 1905:9. A much earlier strike occurred in 1860. L.G. Chafe writes:
"A notable strike occurred in Brigus about 1860 when Capt. Dick Supple led 3,000 men in an attempt to abolish berth money. They were only partly successful. About the 1902 strike Chafe comments: The sealers were now only getting one third of their catch in the steamers, whereas formerly in the sailing vessels they got one half of their voyage. They were also compelled to pay certain charges, which they regarded as unfair and excessive. There were 20 steamers fitted out this spring with about 4,000 men. The majority of these crews fitting out in St. John's decided to make a mighty struggle of labor against capital for better terms of agreement. On March 10th the several crews all united in one great body, and with flags fluttering in the breeze paraded Water Street, marched to the Government House and interviewed His Excellency the Governor, in relation to their grievance. Their demand upon steamer owners was (1) that the price of fat should be $5.00 per quintal; (2) that the charge, $3.00 a man, for coaling and berth money should be abolished; (3) that the 33 per cent increase on supplies issued to the men should be struck off; (4) that food should be provided for men waiting to go on board their respective steamers after a certain given time. After much resistence from the owners of the steamers a compromise was reached. The men were given free berths and were guaranteed $3.50 per cwt. for their seals. (Chafe 1923:38)

THE SEALERS GAINED THE STRIKE
The version given by White (1960) is also found in an unidentified manuscript submitted to MUNFLA by Leo Moakler. In all probability the work is by Johnny Burke.

THE LUCK WENT WITH THE SEALERS SINCE BRAVE COLLOWAY LED THE STRIKE
The version given by White (1960) is also found in an unidentified manuscript submitted to MUNFLA by Leo Moakler. In all probability the work is by Johnny Burke.

ICE BOUND HUNTING SEALS
See footnote citation.

THE SEALER LAD
See footnote citation.

WHEN THE FLIPPERS STRIKE THE TOWN
See footnote citation.

ARRIVAL OF THE *GRAND LAKE* & *VIRGINIA LAKE* WITH BUMPER TRIPS
See footnote citation.

FIRST ARRIVAL *AURORA* & *WALRUS*, FALL
See footnote citation.

ARRIVAL OF *AURORA, DIANA, VIRGINIA LAKE* & *VANGUARD*, LOADED
See footnote citation.

THE *NIMRODS* SONG
For the *Nimrods* Antarctic career see Sir Ernest Shakleton, *HEART OF THE ANTARCTIC* (London, 1911). The largest bill ever made in a St. John's sealing steamer was that made by the crew of the *S.S. Nimrod*, Capt. Peter Cummins, in 1871; crew 140 men, made $208.47 each in two trips — 28,087 seals. (Chafe 1923:36).

CAPTAINS AND SHIPS
Carling 1977:46; Doyle 1955:19; Mowat 1973:48.

THE SEALER'S SONG
See footnote citation.

SIX MEN & ONE WOMAN TAKEN OFF ICE AT PETTY HARBOUR
See footnote citation.

SUCCESS TO THE HARDY SEALERS
See footnote citation.

FIRST ARRIVAL FROM THE SEAL FISHERY
See footnote citation.

CAPTAIN BOB BARTLETT
For a recent account of Captain Bartlett see Horwood 1977.

BALLAD OF CAPTAIN BOB BARTLETT
The same version by Wornell is also given in Smallwood 1967:454. A different version is found in Wakeham 1955:19.

MAURICE CROTTY
Burke and Murphy 1901; Doyle 1927:49; McCawley 1929:46; Mowat 1973:71; Murphy 1905:3; Murphy 1976:155. Michael Murphy (1976) says that this song appeared for the first time in *THE DUKE OF YORK SONGSTER* 1901. A version of this song was collected in 1977, by Hoodie Rawlings, a Folklore Student at Memorial University, from Gerry Bromley of Conche.

THE SPRING OF THE WADHAMS
Dodd: 1972:52. This song is more commonly known as Maurice Crotty. 1852 is generally known and spoken of as the "Spring of the Wadhams." Seals were found very plentiful in the vicinity of the Wadhams, (islands located in Notre Dame Bay S.E. of Fogo Island), and the majority of the vessels were caught in a fearful gale of N.N.E. wind which caused great destruction to the fleet. It was calculated that upwards of 40 vessels were smashed to matchwood by the rafting ice, while more were destroyed by fire by the upsetting of stoves. About 200 men arrived from Greenspond, who reported leaving 1,500 men behind them who only saved what they stood in. The crews were sheltered on the Wadham Island from the 5th to the 12th of April, 'till a relief ship was sent them by the Government (Chafe 1923:41).

I AM A NEWFOUNDLANDER
See footnote citation.

A SEALER'S LOVE LETTER
See footnote citation.

LOSS OF THE *S.S. ALGERINE*
See footnote citation.

IN MEMORIAL
See footnote citation.

THE *NEWFOUNDLAND* DISASTER (1)
The *Newfoundland*, Capt. Wes. Kean, lost 77 men, who were overtaken by a blizzard whilst on the ice (Chafe 1923:44). See also Brown 1974 and Scott 1978.

THE *NEWFOUNDLAND* DISASTER (2)
See footnote citation.

THE *TERRA NOVA*
See footnote citation.

SOUTHERN CROSS
Doyle 1966:41; Doyle 1940:57; Greenleaf 1968:281; Leach 1965:202; Peacock 1965:973; MUNFLA C129, C256, C695, (transcripts); Mowat 1973:135.

CAPT. FREDERICK HARRIS AND THE GRATES COVE SEAL KILLERS OF 1915.
See footnote citation.

AU REVOIR TO OUR HARDY SEALERS
See footnote citation.

THE *KITE* ABANDONED IN WHITE BAY
See footnote citation.

TO THE MEMORY OF THE LATE CAPTAIN KENNEDY
See footnote citation.

THE SWILES OF NEWFOUNDLAND
See footnote citation.

UNCLE BILL TELLER
See footnote citation.

WILLY MARCH
See footnote citation.

THE SEALERS OF NEWFOUNDLAND
See footnote citation.

SUCCESS TO EVERY MAN
See footnote citation.

THE BIRD ROCKS
Blondahl 1964:69; MUNFLA C281. For the same version which Mrs. Walters of Rocky Harbour sung for Greenleaf thirty years earlier — See Peacock 1965:903.

THE COOKS OF TORBAY
See footnote citation.

A NOBLE FLEET OF SEALERS
Bennett 1972:26; Blondahl, 1964:75; Doyle 1966:16; Fowke 1960:16; Mowat 1973:64.

THE LITTLE SWILER
See footnote citation.

THOUGHTS ON THE NEWFOUNDLAND SEALING VOYAGE
See footnote citation.

THE MASTER WATCH
Carroll 1907:12; Carroll 1945:28. See also Blondahl 1964:81; Doyle 1966:68; Johnson 1937:370; Martin 1945:28; Mowat 1973:63.

***NEPTUNE*, RULER OF THE SEA**
G.M. Story states that Captain Edward White supervised the building of the *Neptune*, and had brought her over from Dundee for Job's; and her coming ushered in the last and greatest phase of the hunt...(Story 1962:22).

COTTON'S PATCH (1)
In 1922, Major Sydney Cotton, an Australian aviator, in collaboration with Newfoundlander Sidney Bennett, offered to locate the mythical "Main Patch" or main herd of seals for the local sealing firms. (Mercer 1974:n.p.).

COTTON'S PATCH (2)
Mercer 1974:n.p.

FULL LOADS TO THE SEALERS
See footnote citation.

THE SEALERS' BALL
See footnote citation.

MRS. MULLOWNEY WAS THREE WEEKS IN BED SINCE SHE ATE THE FIPPER STEW
For the same version as White (1960) see fragments of unidentified manuscript submitted to MUNFLA by Leo Moakler. In all probability the work is by Johnny Burke. Courtesy of Paul Mercer. For an eight stanza recipe for flipper pie, see Pratt 1948:40. A two verse 'Flipper song' by M.A. Devine is given in McGrath 1974:5.

THE SEALING CRUISE OF THE *LONE FLIER*
Blondahl 1964:76; Greenleaf 1968:246; Greenleaf 1968:379 (dance tune). Since we feel the Gerald S. Doyle version of this song came from Greenleaf (1968) we include the following note:
"This very frank and spirited modern song is a tune product of group composition. Herbert Watkins said that they all started to make up the song at the beginning of the voyage, and some one added a line or two, or even a verse, whenever anything happened, so that the song is a sort of log of the cruise. When they got back to Twillingate, they put it in the newspaper, the Twillingate Sun, *signing it, "A Young Timer," and omitting as too personal the lines about "donkeying" the man. They did not compose the tune, but chose the tune*

used for "The Lumber Camp Song".
The "Lone Flier" is a motor-auxiliary schooner, owned by the Ashbournes, great merchants of Twillingate. She is used for transporting trade-goods, for sealing in the spring, and at other times, when stocked with goods and fitted with shelves and counter like a floating department store, she is used for trading up and down the East Coast. (Greenleaf 1968:249).

THE SINKING OF THE *NEWFOUNDLAND*
See footnote citation.

THE BARRACKS SONG
See footnote citation.

HUNTING SEALS
Doyle 1927:39; Murphy 1916:44; Murphy 1925:14.

THE FISHERMAN'S SON TO THE ICE IS GONE
Murphy 1905:18.

THE GIRL I LEFT BEHIND ME
Byrnes 1931:149.

THE SEALING FLEET
See footnote citation.

ARCTIC ICE AND FLIPPERS
See footnote citation.

THE OLD TIME SEALERS
See footnote citation.

THE OLD-TIME SEALING FLEET
See footnote citation.

LAST OF THE WOODEN WALLS
See footnote citation.

THE SWILERS' SONG
See footnote citation.

CAPTAIN ABRAM KEAN
"Captain Abram Kean has been fifty-two years master of a ship at the seal fishery. He stands for the greatest number of seals ever brought in by one ship, the greatest weight ever brought in by one ship and the greatest value. The total voyages of his different ships is now well over 800,000 seals, and it is quite possible he will reach the million, as he is still hale and hearty. It is a good thing that there are not many like him or there would be no seals left out there." (Chafe 1923:32).
G.M. Story recollects: *"And I can remember too, on the way to school by street-car, a glimpse of a very old, rosy-cheeked man in immaculate black coat — Captain Abraham Kean, most famous of the seal killers. When he died in 1945, aged ninety, he could boast of having brought back from the ice more than a million seals. And seeing him on his daily visit to the Board of Trade was a visible, and personal, reminder of a long-vanished age — the Homeric age of the Newfoundland seal hunt." (Story 1962:21).*

BALLAD OF SEALING SHIPS AND SEALERS
Wornell 1951:9; Wornell 1955:5; Wornell 1959:27.

THE DYING SEAL-HUNTER
See footnote citation.

THE SEALERS OF TWILLINGATE AND NEW WORLD ISLAND
See footnote citation.

A SEALER'S REPLY TO HIS WIFE
Samson 1962:28.

UNCLE JOHN-THE SEALER, 1951
Samson 1962:22.

WE WILL ALWAYS HAVE OUR SEALERS
Otto Kelland wrote this song on Bowrings wharf during the sealing season of 1945. Five ships were leaving the harbour for the icefields. A gentleman in his eighties who had spent several seasons 'out the ice' said to Kelland, "they'll be the last", in reference to the ships. Kelland disagreed. "What do you know about it," the old man asked? It then occurred to Kelland to write a song but he had no paper. He asked the old man if he had any paper and he took from his pocket an old envelope and gave it to Kelland. Then Kelland told him to get off the gump. He did. Putting the envelope on the gump Kelland then composed, "We Will Always Have Our Sealers." (Eds.)

SAVE OUR SWILERS
Scammell 1977:14.

SONG AND POETRY TITLE INDEX

A Newfoundland Hero 29

A Noble Fleet of Sealers 114

A Sealer's Love Letter 91

A Sealer's Reply to His Wife 153

Arctic Ice and Flippers 137

Arrival of *Aurora, Diana, Virginia Lake* and *Vanguard*, Loaded 73

Arrival of the *Grand Lake* & *Virginia Lake* with Bumper Trips 71

Au Revoir to our Hardy Sealers 102

Ballad of Captain Bob Bartlett 85

Ballad of Sealing Ships and Sealers 147

Barracks Song, The 132

Bird Rocks, The 110

Captain Abram Kean 146

Captain Bill Ryan Left Terry Behind 32

Captains and Ships 76

Captain Bob Bartlett 83

Capt. Frederick Harris 101

Captain Henry Thomey 15

Come All Ye Jolly Ice-Hunters 17

Come Down with the Killock 26

Cooks of Torbay, The 113

Cotton's Patch (1) 120

Cotton's Patch (2) 121

Died on the Ice Fields 39

Driven into Spaniard's Bay 26

Dying Seal-Hunter, The 149

Ferryland Sealer 22

First Arrival from the Seal Fishery 82

First Arrival — *Aurora* & *Walrus*, Full 72

Fisherman's Son to the Ice is Gone, The 134

Full Loads to the Sealers 122

Girl I Left Behind Me, The 134

Greenland Disaster, The (1) 46

Greenland Disaster, The (2) 48

Greenland Disaster, The (3) 50

Greenland Disaster, The (5) 54

Greenland Disaster, The (6) 55

Greenland Disaster, The (7) 57

Hunting Seals 133

I am a Newfoundlander 89

Ice Bound Hunting Seals 67

Ice Floes, The 59

In Memorial of 77 Brave Newfoundland Sealers 93

John Martin, The (2) 21

Kite Abandoned in White Bay, The 103

Land of Fish and Seals, The 25

Last of the Wooden Walls 142

Little Swiler, The 116

Loss of the *Maggie* 41

Loss of the *S.S. Algerine* 92

Luck went with the Sealers since Brave Colloway led the Strike, The 66

Master Watch, The 118

Maurice Crotty 86

Mrs. Mullowney was three weeks in bed since she ate the Fipper Stew 125

Neptune, Ruler of the Sea 119

Newfoundland Disaster, The (1) 94

Newfoundland Disaster, The (2) 96

Nimrods Song, The 74

Old Time Sealers, The 138

Old-Time Sealing Fleet, The 140

Old Time Sealer's Song 18

Save Our Swilers 156

Sealer Lad, The 69

Sealers' Ball, The 123

Sealer's Call, The 14

Sealers Gained the Strike, The 65

Sealers of Newfoundland, The 108

Sealers of Twillingate and New World Island, The 150

Sealers' Song 33

Sealer's Song, The 79

Sealers' Strike of 1902, The 64

Sealing Cruise of the *Lone Flier*, The 126

Sealing Fleet, The 135

Sealing Trip of the *S.S. Greenland* 1891, The 35

Sealing Fifty Years Ago 63

Sinking of the *Newfoundland*, The 130

Six Men and One Woman Taken Off Ice at Petty Harbour 80

Song of the *John Martin* (1) 20

Southern Cross 99

Spring of '97, The 44

Spring of the Wadhams, The 88

Success to every Man 109

Success to the Hardy Sealers 81

Swilers' Song, The 144

Swiles of Newfoundland, The 106

Terra Nova, The 98

Terra Nova Seal Fishing 27

Thoughts on the Newfoundland Sealing Voyage 117

To the Memory of the Late Captain Kennedy 105

Trinity Bay Tragedy 37

Uncle Bill Tiller 106

Uncle John — The Sealer, 1951 154

Wadhams Song 13

We will Always Have Our Sealers 155

We will not go to White Bay with Casey any More 16

When the Flippers Strike the Town 70

Willy March 107

Wreck of the *Maggie,* The 42

Written in Memory of the 48 men lost in *Greenland*
Sealing Disaster March 21, 1898 (4) 53

INDEX TO FIRST LINES

A rugged Newfoundlander as ever sailed the seas 83

A wail from the stricken mothers and wives 53

All pious hearted Christians 54

All welcome to the northern fleet 73

All ye who love old Newfoundland 35

Among the sealers who came home 154

At ten forenoon November five the *Maggie* sailed away 42

Attention! all ye fishermen 65

Attention all, both great and small 101

Attention, boys of Newfoundland 66

Attention, all ye fishermen, and read this ballad down 64

Attention all ye sailor boys 92

And here's grand success to the sealers 122

Bob Bartlett, born in Brigus, of bold sea-faring breed 85

Come all ye friends of Newfoundland 74

Come all you hardy Fishermen 80

Come all ye jolly ice-hunters and listen to my song 17

Come all ye jolly fishermen 20

Come all ye jolly fishermen agoing to the ice 21

Come all ye jolly seal men and listen to my song 126

Come all ye rambling sailor boys 103

Come all ye sons of Newfoundland 96

Come all ye young fellows wherever ye be 113

Come all you Newfoundlanders and listen to my song 156

Come down with the killock 26

Come ye hearty Newfoundlanders, join your voices now with me 147

Dawn from the Foretop! Dawn from the Barrel! 59

De home of his childhood, in Northern Bay 107

De time is drawin' near, my b'ys 109

Dear Miss: — I know I can't mail this 91

Four hundred sail of shipping fine 63

From Bonavista Cape to the Cabot Isles 13

He was such a very little chap 116

Heigho, my lads, for the tenth of March 117

Here Atlantic's foam-wreaths float 142

Ho! We be the Sealers of Newfoundland! 108

I am a Newfoundlander, I go out to the ice 89

I can hear their sirens blowing 149

I must go up to the ice again 14

I think 'twas the spring of the Wadhams 88

I'll go to the ice and I'll catch some seals 134

Let Sunny India her wealth proclaim, her gorgeous glowing sky 25

Mrs. Mullowney gave 125

Newfoundland has many stories that can make the heart beat fast 140

Now that March month has come 153

Oh our schooner and our sloop in Ferryland they do lie 22

Oh, quite early in March, I remember the date 120

Oh the day we left St. John's, me b'ys 106

On Wednesday morning May the 3rd 1944 132

One Monday morning March the Tenth 98

Our gallant ships are going, where rude Boreas is a blowing 102

Sad comes the news from oe'r the sea 93

Sad comes the news from over the sea, from over the troubling main, 50

She got up steam the twelfth of March and shortly did embark 99

Sit down boys, I'll sing you a ditty 86

Slowly today we wended our way 105

Step lively now me bully boys we've got some miles to go 144

Sure when we took 'em to the wharf 123

Terry is a fine young man 32

The bells they are ringing, the sirens are screaming 138

The Block House Flag is up today to welcome home the stranger 33

The fierce wind moaned among the cliffs of rugged Labrador 29

The first arrival from the front 82

The first arrival from the ice 72

The fisherman's son to the ice is gone 134

The *Grand Lake*, boys is coming in 71

The *Neptune*, ruler of the Sea, she rides in court today 119

The *Terra Nova,* Captain Kean 79

The 12th of March is drawing near 81

The white, rugged ice-floe came gliding along 39

The wind was still from the nor' east 67

The sealer lad from his home is gone 69

There's a halo round the margin of the sea 137

There's a noble fleet of sealers 114

They sailed from St. John's Harbour all on the tenth of March 46

Three thousand men of the Viking breed 118

To Harvey's I'll start and to Bowring's I'll go 76

Tom Casey being commander 16

True are the tales which fishermen tell 150

'Twas on the very first day of March 26

'Twas winter down the icy gulf 110

Upon the past I'm thinking 15

Uncle Bill Teller died las' fall 106

We got up steam the ninth of March 121

We have bred many seamen bold 130

We now relate the mournful fate of gallant sealers dead 57

We often read of heroes bold and noble deeds they done 94

We should not forget the Commodore 146

We struck the seals off Cabot Isle 44

We will always have our sealers 155

We'll sound the hardy sealers praise, a wild and cheerful strain 18

What means this hurrying to and fro 135

With knife and fork, with kettle and pan 133

Ye fishermen who know so well 41

Ye hardy sons of Newfoundland, that tread life's rugged way 37

You may talk about the pan-cakes 70

Ye talk o' this, and talk o' that 27

Ye tender hearted Christians, I hope you will attend 48

Present day "landsmen" hunt for seals among the ice floes on Newfoundland's northeast coast.

Photo by Fred Earle, Memorial University Extension Service.

SELECTED BIBLIOGRAPHY

Andrews, C.W. and G.M. **Story**.
William Jackman, DICTIONARY OF CANADIAN BIOGRAPHY X (1972), 378.

Bartlett, Captain Robert.
The Sealing Saga of Newfoundland, NATIONAL GEOGRAPHIC 56:1 (July 1929), 91-130.

Bennett Brewing Company.
NEWFOUNDLAND SONGS. 8th ed. St. John's: Bennett Brewing Co. Ltd., 1973.

Blondahl, Omar.
NEWFOUNDLANDERS, SING! St. John's: E.J. Bonnell Associates, 1964.

Brown, Cassie.
A WINTER'S TALE: THE WRECK OF THE *FLORIZEL*. Toronto: Doubleday Canada Limited, 1976.

Brown, Cassie and Harold **Horwood**.
DEATH ON THE ICE — THE GREAT NEWFOUNDLAND SEALING DISASTER OF 1914. Toronto: Doubleday Canada Limited, 1974.

Burke, John and George **Oliver**.
THE PEOPLE'S SONGSTER, BUYERS' GUIDE & GEMS OF POETRY & PROSE. St. John's: Burke & Oliver, 1900.

Burke, John and James **Murphy**.
THE DUKE OF YORK SONGSTER AND CHRISTMAS ADVERTISER. St. John's: Burke and Murphy, 1901.

Burke, John.
BURKE'S BALLADS. St. John's: John Burke, 1912.

Byrnes, John Mackay.
THE PATHS OF YESTERDAY: MEMORIES OF OLD ST. JOHN'S, NEWFOUNDLAND. Boston: Meador Publishing Co., 1931.

Carling O'Keefe Breweries.
THE NEWFOUNDLAND SONGBOOK. St. John's: Carling O'Keefe Breweries, 1977.

Carroll, Dan.
The Master Watch. THE NEWFOUNDLAND QUARTERLY 7:3 (1907), 12.

Carroll, Dan.
The Master Watch. ATLANTIC GUARDIAN 1:3 (1945), 28.

Chafe, L.G.
CHAFE'S SEALING BOOK. Edited by H.M. Mosdell. St. John's: The Trade Printers and Publishers, Ltd., 1923.

Come all ye Jolly Ice-Hunters. NEWFOUNDLAND STORIES AND BALLADS, 14:2 (1968), 21.

Dodd, Jack.
THE WIND IN THE RIGGING. Torbay, Newfoundland: Jack Dodd, 1972.

Doyle, Gerald S.
OLD TIME SONGS AND POETRY OF NEWFOUNDLAND. St. John's: Gerald S. Doyle Ltd., 1927, 1940, 1955, 1966.

Eckstorm, Fanny and Mary **Smyth**.
MINSTRELSY OF MAINE. Boston: Houghton, Mifflin & Co., 1927.

England, George Allan.
THE GREATEST HUNT IN THE WORLD. Montreal: Tundra Books, 1969. (Originally published as VIKINGS OF THE ICE, 1924).

FAMILY FIRESIDE. St. John's: Gerald S. Doyle.

Fowke, Edith and Alan **Mille** (eds.)
CANADA'S STORY IN SONG. Toronto: W.J. Gage Ltd., 1960.

Fowke, Edith and R. **Johnson** (eds.)
MORE FOLK SONGS OF CANADA. Waterloo, Ontario: Waterloo Music Co. Ltd., 1967.

Fowke, Edith.
THE PENGUIN BOOK OF CANADIAN FOLK SONGS. Middlesex, England: Penguin Books, 1973.

Granger, Charles.
Recent Newfoundland Poetry. In THE BOOK OF NEWFOUNDLAND, J.R. Smallwood, ed. St. John's: Newfoundland Book Publishers Ltd., Vol. 4 (1967), 454-467.

Green, Major W.H.
THE WOODEN WALLS AMONG THE ICE FLOES, TELLING THE ROMANCE OF THE NEWFOUNDLAND SEAL FISHERY. London: Hutchinson and Co. Ltd., 1933.

Greenleaf, Elizabeth (ed.)
BALLADS AND SEA SONGS OF NEWFOUNDLAND. Cambridge: Harvard University Press, 1933; reprinted, Hatboro, Pa.: Folklore Associates 1968.

HARBOUR GRACE STANDARD. Harbour Grace, Newfoundland.

Harrington, Michael.
POEMS OF NEWFOUNDLAND. This volume contains the prize-winning poems in the O'Leary Newfoundland Poetry Award 1944-1952. St. John's: F.M. O'Leary Ltd., 1952.

Horwood, Harold.
BARTLETT, THE GREAT CANADIAN EXPLORER. New York: Doubleday, 1977.

Johnson, Charles H.
Songs and Sagas of Newfoundland. In THE BOOK OF NEWFOUNDLAND, J.R. Smallwood (ed.). St. John's: Newfoundland Book Publishers, Ltd., Vol. II (1937), 219-273.

Kelland, Otto.
ANCHOR WATCH: NEWFOUNDLAND STORIES IN VERSE. St. John's: Dicks & Co., 1960.

King, Clayton L.
THE *VIKINGS* LAST CRUISE. St. John's: Evening Telegram Printers, 1936.

Leach, MacEdward.
FOLK BALLADS AND SONGS OF THE LOWER LABRADOR COAST. Ottawa: National Museum of Canada, 1965.

Loveridge, John C.
STORY IN PICTURES AND POETRY OF THE 1973 SEAL HAUL AT TWILLINGATE NEWFOUNDLAND ALSO MARKING THE 85th YEAR IN BUSINESS OF THE FIRM OF E.J. LINFIELD. Grand Falls: Robinson Blackmore, 1973.

Martin, Ronald (ed.)
POEMS OF ACTION, SENTIMENT AND REFLECTIONS: SELECTIONS FROM NEWFOUNDLAND AUTHORS. St. John's: Privately published, 1945.

McCawley, Stuart.
CAPE BRETON COME—ALL YE: A BOOK OF SONGS AND BALLADS OF CAPE BRETON AND NEWFOUNDLAND. 2nd ed. Glace Bay, N.S.: Brodie Printing Service, 1929.

McGrath, J.W.
The Devine Brothers of Kings Cove — Writers and Balladers. A guest lecture for the Newfoundland Historical Society, St. John's, October 22, 1974.

Mercer, Paul (ed.)
THE BALLADS OF JOHNNY BURKE. St. John's: Newfoundland Historical Society, 1974, Pamphlet No. 1.

Mills, Alan (ed.)
FAVORITE SONGS OF NEWFOUNDLAND. Toronto: BMI Canada Ltd., 1968.

Mowat, Farley.
IN THE WAKE OF THE GREAT SEALERS. Toronto: McClelland & Stewart, 1973.

Murphy, James.
SONGS AND BALLADS OF NEWFOUNDLAND, ANCIENT AND MODERN. St. John's: James Murphy, 1902.

Murphy, James.
SONGS OF OUR LAND: OLD HOME WEEK SOUVENIR. St. John's: Telegram Print, 1904.

Murphy, James.
MURPHY'S SEALER'S SONGBOOK. St. John's: James Murphy, 1905.

Murphy, James.
THE SEAL FISHERY. St. John's: James Murphy, 1916.

Murphy, James.
SONGS OF NEWFOUNDLAND BY VARIOUS AUTHORS. St. John's: James Murphy, 1917.

Murphy, James.
SONGS THEIR FATHERS SUNG. FOR FISHERMEN. OLD TIME DITTIES. St. John's: James Murphy, 1923.

Murphy, James.
SONGS SUNG BY OLD TIME SEALERS OF MANY YEARS AGO. St. John's: James Murphy, 1925.

Murphy, James.
When The Flippers Strike The Town. NEWFOUNDLAND STORIES AND BALLADS, 5:2 (1959), 13.

Murphy, Michael.
PATHWAYS THROUGH YESTERDAY. Gerald Moore, editor. St. John's: Town Crier Publishing Company Ltd., 1976.

OLD HOME WEEK SONGSTER. TERRA NOVA'S KIND WELCOME TO HER SONS AND DAUGHTERS. St. John's: Herald Job Print, 1904.

Peacock, Kenneth.
SONGS OF THE NEWFOUNDLAND OUTPORTS. 3 vols. Ottawa: National Museum of Canada, 1965.

Peddel, Nicholas.
NEWFOUNDLAND POEMS. Harbour Grace: Standard Press, 1904, 1908.

Poetry and Ballads of Newfoundland, in THE BOOK OF NEWFOUNDLAND, J.R. Smallwood (ed.). St. John's: Newfoundland Book Publishers Ltd., Vol. I (1937), 453-486.

Pratt, Art.
BALLY ROT. Devon: Arthur Stockwell Ltd., 1948.

Pratt, E.J.
HERE THE TIDES FLOW. Introduction, Notes and Questions by D.G. Pitt. Toronto: MacMillian Company of Canada Limited, 1962.

Samson, Solomon.
MINUTES OF THE LEISURE HOUR. A COLLECTION OF POEMS, SONGS AND BALLADS PERTAINING TO NEWFOUNDLAND. St. John's: Trade Printers & Publishers, 1962.

Samson, Solomon.
A GLIMPSE OF NEWFOUNDLAND IN POETRY AND PICTURES. Edited by Robert Saunders. Poole, England: J. Looker Ltd., 1963.

Sanger, Chesley W.
Technological and Spatial Adaptation in the Newfoundland Seal Fishery During the Nineteenth Century. (Unpublished M.A. Thesis, Memorial University, 1973).

Scammell, A.R.
SONGS OF A NEWFOUNDLANDER. St. Jerome, P.Q. : J.H. Labelle, 1945.

Scammell, A.R.
Save Our Swilers, DECKS AWASH 6:4 (1977), 6.

Scammell, A.R.
Save Our Swilers, THE EVENING TELEGRAM, June 18, 1977.

Scott, John.
The Function of Folklore in the Interrelationship of the Newfoundland Seal Fishery and the Home Communities of the Sealers, (Unpublished M.A. Thesis, Memorial University, 1974).

Scott, John.
I don't think there's anything in the world that the common man will take a bigger chance for than he'll take for a seal. Some contributions of oral history toward an understanding of the *Newfoundland* Disaster, in Neil V. Rosenberg (ed.), FOLKLORE AND ORAL HISTORY. St. John's: Memorial University of Newfoundland, (1978), 81—90.

Stansford, Joshua.
FIFTY YEARS OF MY LIFE IN NEWFOUNDLAND.
Devon: A.H. Stockwell, 1952.

Story, G.M.
Hon. Edward White: A Great Sealing Captain, THE NEWFOUNDLAND RECORD 1:3 (May — June 1962), 21-22.

Wadhams Song, NEWFOUNDLAND STORIES AND BALLADS 4:1 (1957), 9.

Wakeham, P.J.
King of the Northern Floes, NEWFOUNDLAND STORIES AND BALLADS, 2:3 (1955), 19.

Whelan, Stella.
The Old Time Sealers, NEWFOUNDLAND STORIES AND BALLADS, 10:2 (1964), 33.

White, John.
BURKE'S BALLADS. St. John's: Privately published, 1960.

Wornell, A.C.
RHYMES OF A NEWFOUNDLANDER. St. John's: Guardian Limited, 1951.

Wornell, A.C.
Ballad of Sealing Ships and Sealers,
NEWFOUNDLAND STORIES AND BALLADS, 2:1 (1955), 5.

Wornell, A.C.
Ballad of Sealing Ships and Sealers,
NEWFOUNDLAND QUARTERLY, 58:1 (1959), 37.

Wornell, A.C.
*The **Greenland** Disaster*, NEWFOUNDLAND STORIES AND BALLADS, 8:2 (1962), 27.

Wornell, A.C.
*The **Greenland** Disaster*, NEWFOUNDLAND STORIES AND BALLADS, 19:1 (1972), 31.